MESETA AND CAMPIÑA LANDFORMS IN CENTRAL SPAIN

A Geomorphology of the Alto Henares Basin

Bruce G. Gladfelter

University of Illinois, Chicago Circle

THE UNIVERSITY OF CHICAGO
DEPARTMENT OF GEOGRAPHY
RESEARCH PAPER NO. 130

1971

Library of Congress Catalog Card Number: 75-133028

Research Papers are available from:
The University of Chicago
Department of Geography
5828 S. University Avenue
Chicago, Illinois 60637
Price: $4.50 list; $4.00 series subscription

PREFACE AND ACKNOWLEDGEMENTS

My interest in the fascinating landscape and past of Spain began during graduate course work in prehistoric geography at the University of Wisconsin, Madison. Laboratory training, involving Pleistocene sediments from the Spanish meseta, led to frequent and stimulating discussion of the geomorphology of central Spain. The context and emphasis of these exchanges is reflected by the framework of this study.

The topical focus of the present work is twofold: the Tertiary upland surfaces (påramos) of the Hesperian Meseta, and the Pleistocene record in the broad valleys (campiñas) cut into this plateau. The upland erosional surfaces have received previous attention, but controversy remains regarding the number and evolution of these levels and it is hoped that the data presented in this study will provide further insight as to their origins. On the other hand, apart from detailed studies at the Lower Paleolithic occupation sites of Ambrona and Torralba, virtually nothing is known about Pleistocene materials and morphogenetic environments of the Hesperian Meseta. Consequently, the landforms and depositional record of the campiñas outlined here should permit a more complete understanding of Pleistocene environments and processes in this part of central Spain. These results also provide a potential stratigraphic link between Ambrona and Torralba on the one hand, and the mid-Pleistocene archeological sites of the Manzanares-Jarama valleys near Madrid and the Rio Tajo near Toledo on the other.

The first part of the study is concerned with the contemporary geomorphic setting of the Hesperian Meseta. Structural and bedrock factors are discussed in as far as they are relevant for surface form; climate and the vegetation cover are discussed in their relationships to contemporary geomorphologic process and form. This section concludes with a discussion of the surface form of the Hesperian Meseta. The balance of the study is devoted to the upland erosional surfaces and to Pleistocene features and deposits. The final chapter attempts to relate the alluvial terraces of the study area to those of the lower Henares drainage basin and certain adjacent valleys.

Field work was conducted from July through December, 1967. Extensive traverses and mapping of the upland surfaces were followed by detailed study and mapping of the Pleistocene materials within the lowland campiñas. The second phase of research, spanning the subsequent 17 months, involved data analysis, processing of sediment samples in the laboratory, and formulation of an interim manuscript. The laboratory analysis

iii

was carried out in the Paleo-Ecology laboratory of the University of Chicago. Radiocarbon dates were provided by Isotopes Inc., Westwood, New Jersey. A second, brief, three-week field season during June and July, 1969, allowed final re-evaluation of conclusions. The study was supported by a doctoral dissertation grant (GS-1774) from the National Science Foundation.

Many people have assisted directly or indirectly in the preparation of this manuscript. The field research would not have been initiated without endorsement by the Instituto Geológico y Minero de España, to which I am grateful for cooperation. In Spain, Professor Emiliano Aguirre of the University of Madrid provided invaluable liaison as well as warm hospitality on many occasions. I am also grateful for the interest and hospitality of my friends in Sigüenza (Guadalajara) and Medinaceli (Soria).

I am particularly indebted to the University of Chicago Geography Department for financial assistance throughout the research period, and to students and faculty who provided discussions, criticism or assistance. Daniel Bowman and Claudia Carr gave freely of their time during the laboratory phase of this study. Leslie G. Freeman (Department of Anthropology) kindly examined artifactual materials collected but subsequently not inventoried in this study. Special acknowledgement is due Karl W. Butzer to whom I owe my interest in geomorphology and the Pleistocene. Professor Butzer supervised and guided all phases of this research and his excursions with me in the field in Spain on two separate occasions were rare learning opportunities as well as memorable associations. His patient prodding and exchange of ideas proved of inestimable value. Any inaccurate observations, conclusions, or errors of oversight that may remain, however, are my own responsibility.

My wife, Betsy, shared fully in the burdens of this study. Her participation in the field season, encouragement during formulation of the manuscript, endless typing and editorial talents, and constant support are deeply appreciated.

Bruce G. Gladfelter

Evanston, Illinois
February, 1970

LIST OF TABLES

LIST OF ILLUSTRATIONS

PART I

THE GEOMORPHIC ENVIRONMENT AND SURFACE FORM

CHAPTER I

INTRODUCTION

Much of the interior of the Iberian peninsula comprises a vast, dissected plain known as the Meseta. This expanse of more than 130,000 sq km extends from the Cantabrian piedmont in the north to the footslopes of the Sierra Morena along the southern margin. It terminates in the east at the spine of the Montes Ibéricos, while the western margins of the Meseta abut the intensely folded and faulted Paleozoic rocks of the Iberian Massif. The Cordillera Central comprises an east-west finger of this massif that separates the northern tablelands of Old Castile from those of New Castile to the south.

Tertiary deposits in the Old Castilian Basin form a discontinuous plain that dips gently to the southwest at an average elevation of some 850 m and is moderately dissected by the Rio Duero. Along the northern, eastern, and southern borders, pediment ramps are found adjacent to the Cantabrian, Iberian, and Cordillera Central ranges respectively, with local morphological differences that result from a variance in lithology. However, the Miocene sediments that abut these pediments comprise the major surface expression of the tablelands. Dissection of the Tertiary beds has produced mesas and cuestas, such as in the environs of Valladolid, as well as the more common rolling plains (tierra de campos) with broad valleys and isolated groups of hills.

The expanse of the southern tableland of New Castile owes its more varied relief to the extensive dissection of the Tertiary sediments by the Tajo and Guadiana rivers. The middle course of the Tajo occupies a faulted sedimentary trough at 600 to 700 m between the Cordillera Central and the Montes de Toledo. The latter uplands flank the western margin of the virtually undissected, interior upland plain of La Mancha. The Pontian limestones that mantle this gently dipping upland between 800 and 1000 m comprise the largest plateau of the Iberian Meseta. East and south of Madrid, the upper Tajo and its major tributaries--the Jarama, Henares and Tajuña--have entrenched their courses as much as 200 m into this upland, or páramo, and have formed the buttes, mesas and cuesta forms that are characteristic of most tablelands. The Meseta, defined as a morphologic province, is distinct from geologic provinces since in several areas the tableland surface can be shown to be preserved on the Paleozoic rocks of the Iberian Massif, on Mesozoic strata, as well as on the more widespread Tertiary deposits (see Lautensach and Mayer, 1961).

3

The northern and southern Castilian tablelands are linked by a narrow plateau that forms a depression between the Cordillera Central and the Montes Ibéricos. This saddle has been referred to as the Hesperian Meseta (Lautensach, 1964). The region retains the general morphology of the Castilian tablelands although the features are developed in a somewhat different geologic setting. The Precambrian and Paleozoic rocks of the Iberian Massif have been termed the Hesperian Mass (see Schröder, 1930 and E. Hernández-Pacheco, 1934) and, although the morphologic province considered here is composed of Mesozoic bedrock, the term Hesperian Meseta nevertheless has been applied to the area, presumably because of the Hesperian basement exposed in the Cordillera Central and, to a lesser extent, the Montes Ibéricos. [1]

The Hesperian Meseta forms a tripartite drainage divide between the Duero, Ebro and Tajo. The eastern slopes feed the Rio Ebro system, while drainage to the north is to the Duero Basin. The western flank of the Hesperian Meseta is drained by headwaters of the Rio Tajo, a major tributary of which is the Rio Henares. The areal focus of this study is the Henares watershed that comprises the Henares, Salado and Dulce rivers (see Fig. 1). Defined in this way, the study area is referred to as the Alto Henares, as distinguished from the "upper Henares," which refers only to the immediate Rio Henares valley upstream of the Dulce confluence. Politically, the Alto Henares region falls almost entirely within the northeastern part of the Province of Guadalajara.

The eastern and northern boundaries of the study area are established by the major drainage divides. To the east, this divide is formed by the Sierra de Ministra, which separates the Jalón drainage from the headwaters of the Dulce and the upper Henares. The plateau of Barahona establishes the drainage divide between the Rio Bordecorex in the Duero Basin, and the Salado drainage, so defining the northern boundary of the study area. Streams along the western margin of the Alto Henares in part drain the Somosierra foothills, so that this boundary of the study area is defined by the eastern limit of Paleozoic bedrock of the Cordillera Central. Finally, the southern margins comprise the large Alcarria plateau, which extends southwest to become an integral part of the Tajo Basin tablelands. Since there is no precedent, the "middle Henares" is defined here as that portion of the Henares drainage basin between the Henares-Dulce confluence at Matillas and the Sorbe-Henares confluence at Humanes. The "lower Henares" is that segment of the river reaching downstream from Humanes to the Henares-Jarama confluence near Alcalá de Henares.

---

[1] Schwenzner (1937, 25) defined the southwest boundary of the Hesperian Meseta by a line running through Cuenca, Priego, Baides and Atienza, and the northeast boundary by a line from Somaën, through Berlanga to Ayllón. Defined in this way, the Hesperian Meseta coincides with the Mesozoic bedrock province that intervenes between the Cordillera Central and Iberian Mountain systems.

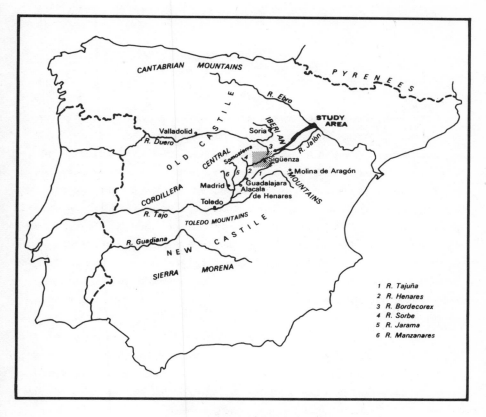

Fig. 1.--Location of the study area

Topographic map coverage of the study area is provided by all or parts of the following sheets of the 1:50,000 Mapa Topográfico Nacional del Instituto Geográfico y Catastral: Barahona (No. 434), Arcos de Jalón (No. 435), Sigüenza (No. 461), Maranchón (No. 462), and Ledanca (No. 487). The latest map editions of this series serve as the reference for place names and physiographic features cited throughout this study, unless noted otherwise.

CHAPTER II

THE GEOLOGIC SETTING

Paleozoic Bedrock

The geologic core of the Iberian Peninsula consists of Paleozoic rocks that are referred to as the Iberian "Mass" or Massif (Lautensach and Mayer, 1961:163). This crystalline complex, strongly folded and deformed during the several phases of the Hercynian orogeny (Solé, 1952:60), supports the surface-form of most of the western half of the Peninsula. In the interior of Spain this crystalline basement is mostly buried in the Ebro, Duero and Tajo Basins by Tertiary sediments that were derived from the Cordillera Central and Cordillera Ibérica, block-faulted and uparched during the Alpine orogeny (Solé, 1952:71). Whereas the Cordillera Central is largely composed of basement rocks, the Cordillera Iberica consists in major part of post-Hercynian strata. The Hesperian Meseta has been planed across a narrow saddle between these two ranges and today provides the only morphologic continuity between the Castilian Mesetas of the Duero and Tajo Basins.

Within the immediate study area, the Paleozoic rocks of the Iberian Massif are exposed in the core of the anticline west of Riba de Santiuste (Fig. 2). Although presumably formed by the oldest of the three major orogenic periods that affected the Iberian Peninsula (Castell and de la Concha, 1956a:26),[1] this anticline was exhumed during the Alpine orogeny (Lotze, 1929:239f) when the overlying Mesozoic rocks were stripped away. The anticline dips at 40° to 60° gently and uniformly to the north but more steeply, or even monoclinally to the south; the strike is N 60°E. Incision by the Rio de Alcolea and its tributaries has exposed a core of Silurian and Devonian strata. Lower Silurian

---

[1]These periods are (see Table 1): (1) the Variscan folding (Hercynian orogeny), late Paleozoic--probably late Carboniferous to Permian in central Spain, (2) Kimmerian orogeny, late Mesozoic--late Triassic to Cretaceous in central Spain, often referred to as the paleo-Alpine orogeny, and (3) the Alpine orogeny, Tertiary and multi-phased. More detailed discussions of those phases of deformation and instability that affected the peripheral environs of the Hesperian Meseta may be found in Lotze (1929,137-298), Richter and Teichmüller (1933,76-100), Schröder (1930,152-78), and Solé (1952,153-204). Except for those few cases where additional comment is required, the discussion in this chapter is restricted to the immediate study area to avoid unnecessary detail.

(Ordovician),[2] fossiliferous quartzites, capped by the thin-bedded, dark blue, fossiliferous shales of Upper Silurian (Gotlandian) age form the core of the anticline. The eastern
end, covering about 4 sq km, comprises yellowish shales with ferruginous sand layers
(early Devonian), lying conformably on the Silurian materials and overlain by Middle
Devonian (Eifelian), blue, fossiliferous, thin-bedded limestones with crystalline structure
and by quartzites (Castell and de la Concha, 1956a:14-15).

In part because of their limited occurrence in the study area, and in part because
of their remoteness within the dissected interior of the Riba de Santiuste anticline, the
Paleozoic rocks are not distinctive in the land-surface form. Erosion of these Paleozoic
shales has produced a hill terrain with forms similar to those developed in the Buntsandstein sandstones (see Fig. 10).

Mesozoic Bedrock

Triassic Sequence

The angular unconformity between the Paleozoic rocks of Riba de Santiuste and the
overlying Triassic sequence attests to a pre-Mesozoic planation of the Iberian Massif.
Schwenzner (1937, 110f) concluded that sufficient equilibrium was obtained to allow the
development of a red weathering horizon upon this surface. The overlying Buntsandstein-
Muschelkalk-Keuper members of the Iberian Triassic system correspond to the typical
germanotype sequence.

The Buntsandstein is locally developed in two members (Castell and de la Concha,
1956a:14-15; 1956b:18-19). The lower unit comprises up to 200 m of semiconsolidated,
poorly stratified, gray-violet, quartzite conglomerate derived from exposed, Paleozoic
uplands; quartz, porphyry and granite components are less common. The upper unit consists of 80 to 100 m of dark red, semi-indurated sandstones, interbedded with shales and
micaceous beds often high in dolomite content. The two units are conformable.

The Buntsandstein characterizes a fluvial or mixed fluvial/littoral detrital facies,
the latter becoming finer in the upper stratigraphic column (Castell and de la Concha,
1956a:15). The gravels are well-rounded but include intact, angular, fractured quartzite
cobbles. Coarse gravel lenses, such as in the Sigüenza area, suggest a fluvial facies
(Lotze, 1929:126).

Exposures of the Buntsandstein within the study area can be observed where erosion has removed the younger members of the Triassic sequence from uparched anticlines.
Protected from planation by the resistant, inclined Muschelkalk that girdles these

---

[2] Ordovician is considered as Lower Silurian in the Spanish and French stratigraphies, rather than as a separate period as in North American stratigraphy.

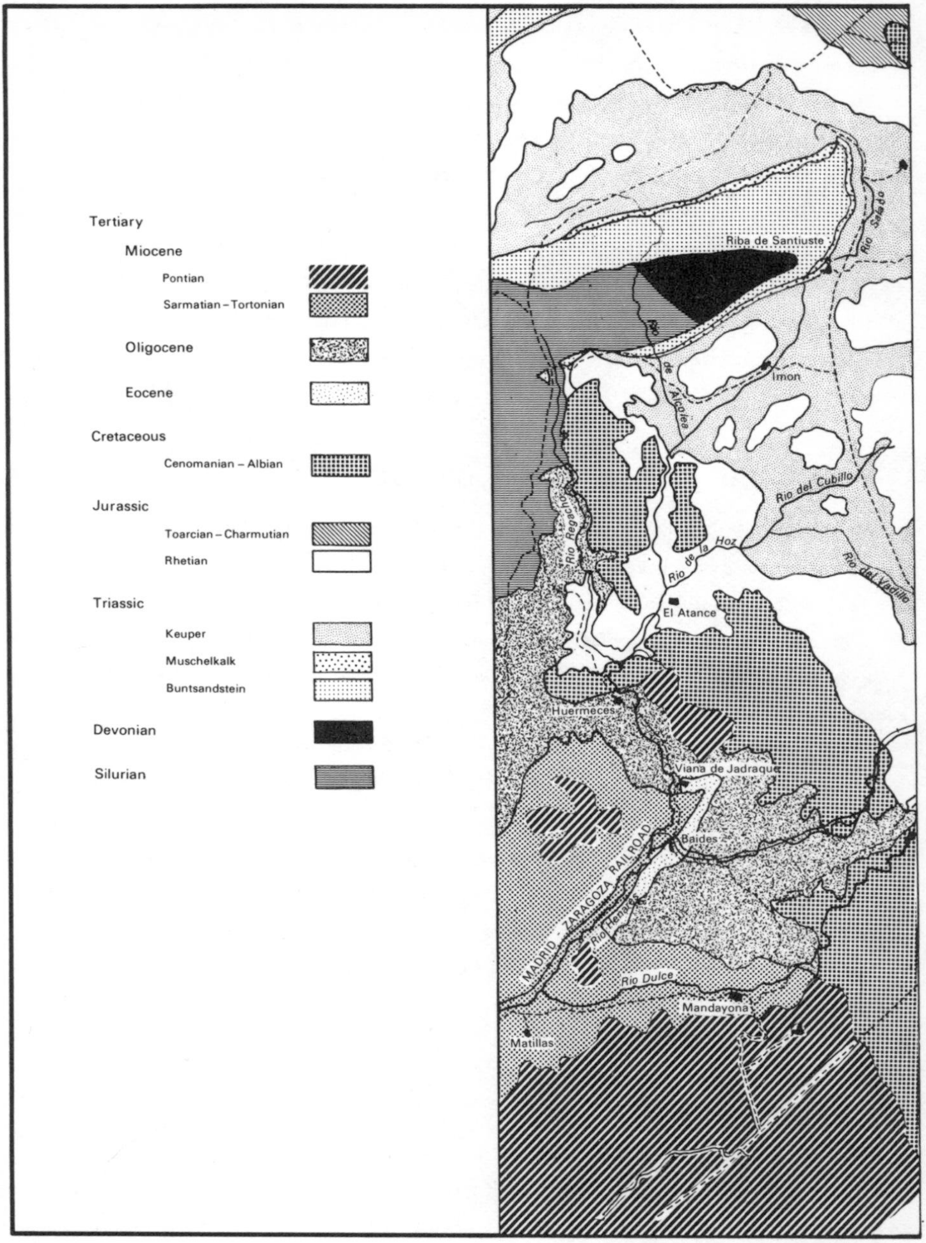

Fig. 2.--Bedrock geology of the study area (after Castell and de la Concha, 1956a, 1956b, de la Concha, 1963, Jordana and Kindelán, 1951, Llopis Lladó and

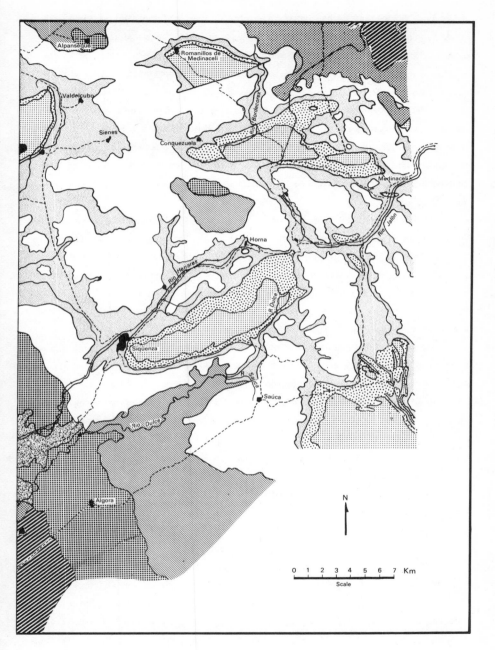

Sánchez de la Torre, 1965, Moya and Kindelán, 1951, Zaranza, 1964, and observations of the author).

structures, fluvial activity has incised and modeled the conglomeratic and sandstone strata to produce a variegated relief that includes pedestal rocks and miniature cuestas preserved by conglomerate caps. Greater local relief (10 m per 100 m sq) is provided by conglomeratic knolls set above concave or flat-bottomed, low-order drainage lines now choked with fluvial/colluvial red sands. These areas, classified as "hills" (Fig. 10), occur most prominently within the Riba de Santiuste and Sigüenza anticlines, as well as on a smaller scale across the upland surfaces northeast of the study area, to the east of Romanillos de Medinaceli and Conquezuela. The variety of land-surface form provided by the Buntsandstein to the otherwise repetitious campiña-páramo of the Hesperian Meseta landscape is further accentuated by the woodlands of pine and oak found on this sandstone (see Chap. III).

The compact, thin-bedded, white to gray, dolomitic limestones of the Muschelkalk are conformable with the Buntsandstein. Rarely more than 30 m thick, this fossiliferous limestone was deposited during a regression of the Triassic seas (Castell and de la Concha, 1956a:15). Some authors recognize a more sandy upper unit of the Muschelkalk intercalcated with greenish marls, marly limestone, calcarenite and micaceous sands (Lotze, 1929:124ff; Schröder, 1930:135-39). Such a facies can be recognized near Cincovillas and within strata exposed by the quarry east of Horna.

The Muschelkalk limestone is exposed on the flanks of anticlines steeply inclined between 30° and 45°; in some cases, such as the 1180 to 1200 m upland west of Pico de Ministra, the deformed limestone appears monoclinal. Even though very thin in development, these inclined Muschelkalk beds were resistant to the erosion that stripped off the overlying Keuper, so that the surface expression of the Muschelkalk assumes a prominence out of all proportion to its thickness. Steep inclinations favor rapid runoff that inhibits pedogenesis; as a result exposed Muschelkalk rocks of the anticlines are barren of vegetation. Where channelled runoff from the weaker anticlinal cores has breached the Muschelkalk, deep "V"-shaped chevron incisions, several hundred meters wide, have been cut. Prominent examples are found northeast of Sigüenza (Arroyo de la Calera), at Guijosa (Rio Quinto), at Barahona (Arroyo de los Tejares), and at Riba de Santiuste. Additional attention is drawn to the Muschelkalk outcrops by the agglomerated settlements frequently found on these resistant and stable exposures. Sigüenza, Alcolea del Pinar, Guijosa, the Castillo at Riba de Santiuste, Rienda, and Cincovillas are examples.

The Keuper shales that complete the upper 200 m of the Triassic sequence are referred to as "variegated marls" by Spanish authors (Castell and de la Concha, 1956a:16; 1956b:20). Schröder (1930, 139-41) was able to recognize three vertical facies within the Keuper but the complexity of the vertical and horizontal changes of this unit in the study area precludes any meaningful stratigraphic delineation. The predominant facies include massive-bedded, white to gray-blue, gypsiferous shales and little-consolidated, plastic, red shales or clays with quartz crystals (<u>Jacintos de Compostela</u>) and aragonite inclusions. Local color changes, involving yellow and black hues, further characterize these <u>margas irisadas</u>. Keuper fossils have been reported (Jordana, 1935:39) but as yet none have been identified within the study area. Sodium chloride, precipitated from mineral-rich waters that percolate through the Keuper, provides the basis for the local commercial saltworks at Salinas de Medinaceli, Alcuneza, Salinas de la Olmedilla, Imón, Salinas La Victoria, and Paredes de Sigüenza.

Keuper units are characteristically found along the concave midslopes and foot-slopes of the valley bottoms of the Mesozoic Hesperian Meseta. The loosely consolidated shales readily yield to fluvial dissection and denudation so that the modern Keuper cam-piñas typically are smoothly undulating, with slopes rarely greater than 25°. Local relief is in excess of 25 to 40 m only where the Keuper is preserved by more resistant caps of Mesozoic or Pleistocene materials. The gently concave (7°-24°) lower slopes bordering these campiñas are prone to mass movement when the impermeable Keuper clayey shales become saturated. Even today massive earthflow scars indicate that undermining of the upland limestones eventually will lead to collapse of huge limestone masses that contrib-ute to the crude detritus of the lower Keuper slopes. This activity suggests that a form of backwearing has been prominent in destroying the páramo surfaces of the Hesperian Meseta. Where steeply embanked against the dip slope of the Muschelkalk, such as east of Sigüenza and north of Riba de Santiuste, scalloped gray Keuper shale and white marly limestones form chevron hogbacks in the order of 100 m wide. These morphologic fea-tures are distinctive amid the otherwise amorphous Keuper land-surface form.

The complete Triassic sequence is best exposed by the dismantled anticline east of Sigüenza. According to Jordana and Kindelán (1951, 53f), this deformation occurred during the late Triassic as a result of movements associated with the early Kimmerian orogeny. Presumably the Triassic-Liassic unconformities are related to this same deformation.

Jurassic Sequence

Between the typical germanotype Triassic sequence and the marine-epicontinental Liassic limestones and sands there are some 200 to 250 m of dolomitic limestone. The time-stratigraphic position of this non-fossiliferous unit is unclear. Spanish authors refer to this transgressive marine sediment as carniolas (de Novo and Chicarro, 1957: 1410; Solé, 1952:85), and relegate it to the Rhetian stage which, in the tradition of French authors, is usually held to be the lowermost unit of the Liassic (see Castell and de la Con-cha, 1956a:17; 1956b:21; 1959:18; Sánchez de la Torre, 1963:128). German authors classify the Rhetian as the uppermost Triassic phase (Brinkman, 1960:Table facing p. 72; Schröder, 1930:142).[3] It seems, therefore, that the Rhetian sedimentation ranged from terminal Triassic to infra-Liassic (Schröder, 1930:142; Castell and de la Concha, 1956a: 17), so that the carniolas limestone will be referred to here simply as Rhetian.

The light gray, dolomitic, carniolas limestone, locally cavernous and commonly without conspicuous bedding, is prominent in the landscape as a resistant caprock over the Keuper, preserving the high páramo surfaces. Warping frequently can be observed in response to readjustments of the underlying, lubricated Keuper clay (see, for example, Llopis Lladó and Sánchez de la Torre, 1965). Where incised by major drainage lines, subangular crestslopes are the rule. Locally, microlenses and flecks of silicates may aid in distinguishing this limestone from the widespread fossiliferous Liassic limestones.

Massive-bedded, gray to dark blue, compact, locally crystalline limestones, marls and sands of the Upper Liassic overlie the carniolas (Moya and Kindelán, 1951:34; Jordana, 1935:40ff). The limestone varies in color (yellow, pink, or black) as well as texture (occasionally oölitic) but is generally rich in fossils. The marls are equally varied in color (gray-blue, reddish) and often clayey in composition, with large quantities of fossils. Whitish or yellow and red, coarse quartz, quartzite and lime sands occur in the uppermost Liassic unit (Moya and Kindelán, 1951:34). The thickness of this entire Upper Liassic (Charmutian-Toarcian) sequence is less than 100 m.

Topographic expression is confined to gently undulating uplands preserved in the limestone unit. Smooth-floored valley depressions with limited local relief are supported by the upper unit of sands and loosely consolidated sandstones (Fig. 10).

Cretaceous Sequence

The town of Barahona stands on a knoll of Cretaceous limestone that rests discon-formably on the Upper Liassic materials of the upland páramo surface (Castell and de la Concha, 1956:18 and geologic profiles). Although no precise analyses have been made,

---

[3]Jordana and Kindelán (1951) avoid any such confusion by omitting the carniolas limestone from their treatment of the geology of the Sigüenza area. Presumably because of the absence of paleontological confirmation, these authors consider much of the limestone cap of the 1:50,000 Sigüenza geologic map to be Lower Cretaceous (Albian) in age rather than Rhetian. This interpretation is based on questionable minerological and stratigraphical arguments without paleontological evidence (1951,38f). With the exception of the adjoining Ledanca geologic sheet, which is co-authored by Kindelán, a Lower Cretaceous (Albian) limestone is not recorded elsewhere in the Hesperian Meseta study area. Additional criticisms of the work of these authors need not be enumerated, although the imposition of Pliocene limestone on top of Pleistocene alluvium is too blatant to overlook (Moya and Kindelán, 1951:Geologic profile III-III). Because of these inaccuracies, little reliance has been placed upon these studies.

most authors recognize in this interruption of the Mesozoic sedimentation a record of a
paleo-Alpine period of folding known as the Kimmerian orogeny (Lotze, 1929:148f; Schröder,
1930:152; Jordana and Kindelán, 1951:54f; Moya and Kindelán, 1951:26; Solé, 1952:89;
Castell and de la Concha, 1956a:21; 1956b:35; 1959:26f). It should be noted that Sánchez
de la Torre's (1963) detailed and valuable study of the northeastern periphery of the Hesperian Meseta does not contradict this notion but rather recognizes a local distinction:
"El Lías y el Cretácico se encuentran concordantes, por lo que no fue afectada esta zona
[italics mine] por los plegamientos kiméricos ... " (p. 131). The moderate folding of
this period was largely occluded by the intense deformation of the subsequent Tertiary
orogeny but there is no question that the post-Liassic--pre-Albian hiatus recorded in the
Hesperian Meseta was a period of emergence, and it would seem that sufficient equilibrium obtained to allow a white, kaolinite weathering product to develop on the pre-
Cenomanian erosional surface (Schwenzner, 1937:111).

Within the study area, the Cretaceous sequence is recorded along the southwest
margin of the Mesozoic saddle where it has been uparched by the Alpine orogeny. This
area has been mapped by the Instituto Geológico y Minero de España but unfortunately, as
previously indicated, the two relevant memoirs are of dubious value.[4] Description of the
Cretaceous sequence is, therefore, best drawn from alternative sources (e.g., Castell
and de la Concha, 1956:18).

The Lower Cretaceous is recorded by the non-fossiliferous sands and loosely consolidated sandstones of the Albian stage. These whitish, fine sands consist mainly of
quartz with occasional feldspar and quartzite grains. Local thicknesses of 10 to 12 m are
recorded on the Barahona upland where the overlying Cenomanian cap is intact, but maximum thickness is assumed to be greater since 60 to 80 m of this unit are exposed northeast of the drainage divide (Castell and de la Concha, 1959:20) and west of Sigüenza at the
foot of the Sierra Guadarrama (Schröder, 1930:143). Upper Cretaceous (Cenomanian)
limestones lie conformably above the Albian sands. The fossiliferous, thin-bedded limestone is interbedded with thin layers of oxidized marls and occasional chert nodules
(Schröder, 1930:144); overall thickness is estimated to exceed 60 m. An additional Upper
Cretaceous (Senonian-Turtonian) sequence of some 200 m of massive-bedded, fossiliferous limestones is recognized by Schröder (1930,145f; 172-75) exposed among the folds of
the Mesozoic-Tertiary hinge line. This facies is not recorded in that area by other
authors or elsewhere in the study area or Guadalajara Province.

The Cretaceous sequence enjoys only limited morphological expression within the
study area. A gently-rounded Cenomanian knoll rises less than 40 m above the

[4]The 1:50,000 Sigüenza geologic map is currently being remapped at 1:25,000 but
only two of the six maps proposed were available at the time of writing. These are in preliminary form.

general páramo surface at the town of Barahona while less prominent remnant caps occur at Alpanseque and La Ventosa. The most prominent expression is the residual rising more than 100 m above the Alcarria páramo surface north of Algora, the Loma San Cristobal (1213 m).

Cenozoic Bedrock

Early Tertiary

Paleogene

The Upper Cretaceous-Lower Tertiary sedimentary hiatus occurring generally throughout the Meseta is a consequence of regression of the Cretaceous seas that deposited the Cenomanian limestones on the continental Albian sands (Solé, 1952:91-94, Figs. 29,30). It is likely that this recession was accompanied by isostatic adjustment and, on the basis of evidence from elsewhere along the Meseta margins, some limited orogenic movement, perhaps associated with echos of the major Laramidian orogenic phase (see Moya and Kindelán, 1951:54; Solé, 1952:91-94, Figs. 29,30; Crusafont et al., 1960:244f). Subaerial erosion of these Mesozoic upland surfaces led to deposition of a lagoonal facies in adjoining basins. This is best recorded by the steeply inclined Eocene beds now exposed along the left bank of the lower Rio Salado at Huérmeces and Viana de Jadraque. Here monoclinal, white limestones, marls and calcareous breccias dip 70° at the edge of the Cenozoic Tajo Basin, resting with an erosional disconformity against the Upper Cretaceous limestones. These strata are rich in Upper Eocene (Ludian) mammalian fauna and freshwater gastropods (see Schröder, 1930:147f, and Crusafont et al., 1960:248f). Although the Eocene beds do not appear elsewhere within the Hesperian Meseta or the Henares Basin, it is assumed that they are probably found under younger deposits through much of the Cenozoic Tajo Basin (de la Concha, 1963:15).

Sedimentation continued during the Oligocene but it is not clear whether or not these deposits are separated from the Eocene beds by an erosional disconformity and whether an intervening period of uplift can be inferred (see Schröder, 1930:148; Crusafont et al., 1960:250, Figs. 3,7). Two units generally are identified within the Oligocene but their conformity indicates a change of depositional environment rather than an interruption of sedimentation. The distinction is greater at the border of the Henares basin than in the center and identification of these units must be made lithologically (de la Concha, 1963:14).

The bottom facies of early Oligocene age comprises red marls intercalcated with massive gypsum lenses and thin layers of conglomerates, coarse sands, limestone and red to gray-green clays. A Sannoisian molluscan fauna has been identified in compact, yellowish-red limestone in the Rio Cañamares (Schröder, 1930:148) and Rio Aliendre valleys (de la Concha, 1963:20).

The later Oligocene unit includes thick conglomerate beds composed of rounded gravels (2–4 cm in major axis) of quartzite, quartz and limestone, cemented in a calcareous matrix. These alternate with coarse red sandstones or clay bands that grade laterally into beds of sandy marls, marls or limestone (Jordana, 1935:48–49; de la Concha, 1963: 15). In the absence of an erosional disconformity or faunal evidence, it may be more appropriate to refer to Oligocene vertical facies changes rather than Lower and Upper units. Nonetheless, de la Concha (1963, 15) has observed distinctive units on the right bank of the Rio Henares and Riba (1957, 6–8) refers to the early Oligocene unit as the Cogolludo Facies and the later Oligocene series as the Alcarria Facies.

According to de la Concha (1963, 15), the combined thickness of the Oligocene sequence exceeds 1000 m, suggesting a protracted and/or vigorous period of upland denudation involving the stripping of a soil mantle that gave the characteristic reddish hue to these sediments.

Mid-Tertiary Orogeny

The Alpine paroxism that so radically altered the morphology of the Iberian Peninsula by folding, faulting and uparching what today are the Pyrenees and the Cantabrian and Central Cordilleras, did not occur synchronously nor with the same intensity in all of these provinces (see Solé, 1952:Chap. 5). The complexity of this system is further represented by the fact that the mid-Tertiary orogeny comprised multiple phases or waves that were not completely recorded throughout the peninsula. As a consequence of this areal variation and temporal complexity, all authorities are not in agreement in regard to the details of the neo-Alpine orogeny within central Spain.

Most authors do agree, however, that the major period of deformation within the Hesperian Meseta had commenced by late Oligocene times (see Solé, 1952:171f, 188ff; Crusafont et al., 1960:251f) and that, for the most part, it had terminated by mid-Miocene times (Castell and de la Concha, 1952a:27). There is evidence to this effect in the headwater margins of the major Tertiary Basins. In the middle Henares region the deformed and faulted early and late Oligocene beds (see de la Concha, 1963:24 and profiles) show that the paroxism postdated the major period of Oligocene sedimentation. The early Miocene (Aquitanian, Burdigalian) break in the sedimentary record noted by Crusafont et al. (1960, 252) at the Tertiary-Mesozoic contact along the lower Rio Salado may be cited as evidence of orogenic activity. Furthermore, angular disconformities between the Oligocene beds and the younger Miocene sequence are recorded in the Alto Henares between Humanes and Matillas, at La Cabrera in the Rio Dulce gorge, and between Pinilla de

Jadraque and Jadraque in the Rio Cañamares valley (see also de la Concha, 1963:16 and profiles). The essentially undisturbed Miocene beds have not experienced extensive deformation so that the major mid-Tertiary orogeny can be presumed to have ended by the mid-Miocene (Vĭndobonian or Helvetian). This paroxism is believed to have occurred in several spasms, but the principal phase ended by early Miocene (pre-Burdigalian) times (Crusafont et al., 1960:251). Spanish authors refer to this orogenic phase as the Stampian (Estampiense), which corresponds to the Savian phase of H. Stille and his school (Crusafont et al., 1960: 251, 252; see also Moya and Kindelán, 1951:41). It is thought that Savian spasms, continuing into the subsequent Styrian phase, created the extensive Miocene basins that now abut the Hesperian Meseta, and that a later Styrian phase was responsible for the gentle folding of the Tortonian gypsiferous clays (Jordana and Kindelán, 1951: 55; Moya and Kindelán, 1951:41, 54f; Solé, 1952:171f; Castell and de la Concha, 1956a:27).

The Hesperian Meseta experienced strong uparching during the mid-Tertiary orogeny. Sandwiched between the Tertiary block-faults in the Paleozoic core of the Cordillera Central and the synclinal folds of the thick Mesozoic deposits of the Cordillera Iberica, the Mesozoic rocks of the study area experienced warping, folding and faulting in response to the lateral pressures of these adjacent geologic provinces. The resultant northnorthwest to south-southeast strike of the Hesperian folds is reflected in the northwest-southeast trend of the headwaters of the Henares, Tajuña and Jalón rivers and their respective tributaries (see Solé, 1952:310-28, Fig. 127). The strike of the folds is further evident in the warped Cretaceous limestones of the upland surfaces northwest and southeast of Algora (see also Schwenzner, 1937:Map 2). Where ancillary faulting occurred in the otherwise pliable Mesozoic materials, a similar northwest-southeast strike pattern also developed. Though largely concealed by younger planation episodes, these fault scarps are occasionally preserved in the contemporary surface morphology as local cuesta escarpments such as east of Conquezuela and east of Garbajosa.

A zone of maximum deformation is recorded along the southwest margin of the Hesperian saddle. At La Cabrera and again at Huérmeces, Oligocene sediments of the Tertiary Tajo Basin margin are upwarped where they rest with an erosional disconformity against steeply dipping Cretaceous limestone. This contact zone is the area of maximum mid-Tertiary deformation and forms a hinge-line that defines the southern and western borders of the Hesperian Meseta.

In overview, the radical transformations of surface form wrought by the mid-Tertiary orogeny continue to exist in subdued form in the present morphology of the Hesperian Meseta. Broad, undulating swells of the upland páramo surfaces reflect in part on the impact of this orogeny and in part on the plasticity of the Keuper unit underlying the Mesozoic limestones that comprise these surfaces.

Late Tertiary Sequence

Miocene

The large geosynclinal or block-faulted depressions of interior Spain were created or deepened by the mid-Tertiary orogeny, becoming centers of Miocene sedimentation. Throughout the Duero, Tajo, and Ebro Basins the Miocene sequences share broad similarities in the three late Miocene sedimentary units that are generally recognized (see Lotze, 1929: 127-36; Solé, 1952:124; Crusafont and Truyols, 1960). This vertical succession was first worked out for the Jalón sector of the Ebro Basin (Lotze, 1929:132), but the same overall sequence has also been recognized in the Tajo Basin (Schröder, 1930:149ff; Schwenzner, 1937:41; Riba, 1957:5ff). The facies includes (from bottom to top): (1) terrestrial beds, (2) mixed terrestrial and lacustrine beds, (3) a lacustrine sequence. These comprise basal sands, conglomerates and clays, intermediate sands, clays and precipitates, and a terminal chalky limestone respectively.[5] The internal conformities as well as the frequent horizontal changes of facies compound the complexity of the sedimentary record.

The Miocene sequence within the study area is best recorded along the middle course of the Rio Henares in the vicinity of Jadraque. Here the lower facies is comprised of massive, loosely consolidated quartzite, quartz and limestone conglomerates in a reddish-brown to brownish-yellow, sandy or clayey marl matrix. The conglomerates may occur as massive strata up to 10 m in thickness, as above Mandayona, or as the thin, 1 to 2 m lenses intercalated with the red, clayey marl as found in the Jadraque area. While paleontological evidence is sparse (Dantin, 1924), the approximately 70 m column of this facies is ascribed to the Tortonian (Schwenzner, 1937:41; Castell and de la Concha, 1959:21).

The intermediate succession of the Miocene sequence comprises up to 150 m of variegated, sandy red marls (Schwenzner, 1937:41). These vary in hue from green to reddish-brown or yellow, and in structure from compacted granular to laminar. Vertically as well as horizontally the gray or reddish limy clay components of the marl vary as do the fine-grained, reddish quartz and limestone sands (Jordana, 1935:51). This marly unit is interbedded with multi-colored, gypsiferous layers and loosely consolidated conglomeratic lenses that vary between 0.6 and 1.2 m in thickness (de la Concha, 1963:16). Sarmatian fauna have been recovered from this sandy clay-marl (Jordana, 1935:53).

Spanish authors commonly refer to the middle Miocene, that is to the Helvetian-Tortonian-Sarmatian facies, as Víndobonian which, in the absence of faunal or disconformal

---

[5]Royo Gómez (1928) classifies the Miocene sequence as consisting of riverine deposits (Tortonian), brackish pond deposits (Sarmatian) and swamp deposits (Pontian) respectively.

evidence, often is a convenience that obviates the need for a Tortonian-Sarmatian boundary. De la Concha (1963, 16) describes 140 m of Víndobonian beds immediately west of the study area.

The youngest Miocene unit is composed of as much as 60 m of Pontian limestone,[6] a massive-bedded, white, marly limestone. The nearly horizontal strata individually range from 0.5 to 1.0 m in thickness but may attain 3 m. Although light hues and chromas predominate, shades of gray, yellow, rose or blue locally emerge on fresh surfaces; weathering imparts a reddish-brown cast to long-exposed surfaces. The fossiliferous Pontian limestone is fine-grained, sometimes oölitic, and given to conchoidal fracturing where silica abounds (Jordana, 1935:51). Chalcedonic flint and cherty flint are also found in the limestone.

The variety of facies of the Miocene sequence relates to the original derivation of these sediments from the upland rims of the drainage divides. The several facies record relative depositional situations with respect to the center and peripheries of each basin. These differences perhaps more than any other factor account for the vertical and horizontal complexity of the Miocene deposits. So, for example, it can be shown that in the lower Henares Basin, chemical precipitates were deposited contemporaneously with the detrital sediments that dominate the Víndobonian sequence in the study area (Schröder, 1930:150; Schwenzner, 1937:41; Riba, 1957:6ff; see also Sánchez de la Torre's [1963] detailed analysis of the Miocene strata northeast of the study area).

In addition to the variable character of sedimentation, Miocene deformation was subject to areal variation. Oligo-Miocene conformity has been noted near the center of the Tajo Basin (Riba, 1957:8; Crusafont et al., 1960:250) whereas disconformities can be documented in the middle Henares (de la Concha, 1963:16). Consequently, it would appear that where mid-Tertiary uplift and deformation were most severe, sedimentation was interrupted by intervals of active erosion; however, toward the center of the Tajo Basin sedimentation seems to have continued with little interruption. Solé (1952, 125f) cites a comparable situation for the Víndobonian and Pontian of the Ebro Basin. Consequently, it may be inferred that deformation perhaps occurred throughout the Miocene as a result of isostatic adjustment[7] whereby continuous potential energy was provided for upland denudation and Miocene basin sedimentation (see Solé, 1952:134-38; Castell and de la Concha, 1956a:27).

---

[6] German geologists working in Spain have followed the Spanish tradition of including the Pontian in the Miocene; continental stratigraphers generally designate the Pontian as Lower Pliocene. The former procedure will be followed here as a matter of convenience.

[7] This may also be implied by Birot and Solé (1954, 48): " ... no conocemos período de estabilidad importante en el Mioceno. "

The morphological significance of the Oligo-Miocene strata is conspicuous in the landscape. In contradistinction to the general occurrence of comparatively gentle breaks of slope that characterize the Mesozoic countryside, the Tertiary areas portray an association of surface forms strongly reminiscent of semiarid tableland scenery. Most prominent are the broad páramos, such as La Alcarria, capped by Pontian limestones. This extensive, nearly flat, unbroken upland surface gives way abruptly to angular crest slopes that define the meseta edge. Steep, straight-sided midslopes prevail, preserved in delicate equilibrium among the weaker sands, clays, and marls of the underlying Miocene or Oligocene beds. The acuity of the landscape is reinforced by the color schemes of these facies; reddish-yellow, or seemingly orange, Víndobonian sediments are intermittently broken by whitish layers of gypsum or marly strata that accentuate the horizontal dimension of the landscape. Occasionally the valley bottoms or campiñas are interrupted by mesas that portray this colorful sequence. Where massive Oligocene or Miocene conglomerates have offered more resistance to retreating midslopes, esplanades are locally developed. Elsewhere the protective limestone caps have been destroyed to expose the weaker Oligo-Miocene sand-clay-marl associations and dissection and denudation have sculptured "badland" terrain.

The campiñas of the Tertiary countryside are open and flat-bottomed. Modeled by broad, meandering streams during the course of the Pleistocene (see Chap. VI), the margins of the intensively cultivated valleys break in an inverted subangular fashion to give way to flat-topped remnants of fluvial terraces or directly to the midslopes of the upland surfaces.

Pliocene

S. de la Concha (1963, 17) tentatively identifies subrounded quartzite gravels and sands that occur on the Aliendre-Bornova interfluve (940-980 m) as Pliocene detritus. These gravels overlie the Miocene sequence but Pleistocene derivation is supposedly precluded by the fact that their occurrence is independent of the present drainage network. Origins may relate to a late Pliocene erosional cycle that followed limited deformations associated with the Rhodanian phase of the Alpine orogeny (de la Concha, 1963:25). Correlation with the well-known but controversial raña deposits of the Castilian Mesetas is suggested, but firm conclusions must be reserved (Riba, 1957:18 and references).

Summary

The foregoing discussion of the geology of the study area establishes the structural setting of the geomorphology of the Hesperian Meseta. The lithology and structure of the

TABLE 1

GEOLOGY OF THE STUDY AREA

Geologic Column					Events and Characteristics
Era	Period and System	Epoch and Series	Age and Stage (as present locally)	Orogenies Recorded	Lithology, Depositional Environments*
Cenozoic	Tertiary	Pliocene	. . . . . .	Rhodanian	Gravel veneer (rañas?); terrestrial ⌇ disconformity ⌇ Limestone; continental, @ ⌇
		Miocene	Pontian — Sarmatian, Tortonian, Helvetian (Vindobonian)	Styrian / Savian	Sandy marls, clays, sands; gypsum; terrestrial ⌇ Conglomerates, sands; gypsiferous marls and clays; lacustrine/mixed terrestrial, @ ⌇ (no deposits) disconformity ⌇ Gypsiferous marls, conglomerate, sands; clays and limestone; continental, @ ⌇
		Oligocene	Sannoisian		disconformity ⌇
		Eocene	Ludian		Limestone; continental, @ ⌇ (no deposits) disconformity ⌇
		Paleocene		Laramidian	
	Cretaceous	Upper	Cenomanian	Laramidian	Limestone; marine, @ ⌇ disconformity ⌇
		Lower	Albian	Late Kimmerian	Sands, loosely consolidated sandstone; epicontinental ⌇ disconformity ⌇

(Alpine — spanning the Orogenies Recorded column)

Era	System	Series	Stage	Phase	Orogeny	Lithology
Mesozoic	Jurassic	Liassic	Toarcian / Charmutian	Early Kimmerian	— Kimmerian —	Limestone, marls, sands; marine/epicontinental, @
	Triassic	Keuper	Rhetian			Limestone; marine ～～ disconformity ～～ Shale; marine, @
			Carnian/Norian			
		Muschelkalk	Anisian/Ladinian			Limestone; continental, @
		Buntsandstein	Scythian	Variscan		Conglomerate and sandstone; littoral/continental ～～ disconformity ～～ Thin bedded limestones, quartzites; marine
Paleozoic	Devonian	Middle	Eifelian		— Hercynian —	
		Lower	?			Shales and sands; marine
	Silurian	Upper	Gotlandian			Shales; marine @
		Lower	Ordovician			Quartzite; marine @

*Fossiliferous beds are indicated by @.

bedrock have been described and the role of individual bedrock units in defining land-surface form has been briefly considered. A résumé of the geologic units and the orogenic phases recorded in the study area is presented in Table 1. Non-geologic factors relevant to contemporary geomorphic activity are considered in the following chapter.

CHAPTER III

THE BIO-CLIMATIC SETTING AND CONTEMPORARY

GEOMORPHIC ACTIVITY

Climate of the Hesperian Meseta

The limited variability of daily weather on the Spanish Meseta and gradual seasonal transitions, unmarked by dramatic changes in the botanic cycle, contribute to an impression of climatic monotony. Even though parameters of climate disclose significant seasonal differences, these contrasts are less sensible, when experienced in a temporal continuum, than the data might suggest. Summer is long and hot; limited relief derives from low levels of daily relative humidity and the nearly constant breeze or wind that sweeps the unbroken surfaces of the Meseta upland. Cloudless skies promote nocturnal radiant cooling that offers a pleasant relief from the mid-day heat. In absolute terms, Meseta winters are moderate, although when contrasted with summer temperatures, their severity seems accentuated. Mean daily winter temperatures rise considerably above the freezing point, but higher relative humidities, persistent winds, and a high incidence of overcast or cloudy skies intensify the subjective feeling of cold. In actuality, the mean annual temperature range is less than the mean diurnal temperature range of the warmest summer month.

Moisture factors, rather than temperature, assume the dominant sensible roles in distinguishing the transitional seasons of spring and autumn. The primary annual precipitation maximum, occurring in late May or early June, and the late autumn secondary maximum, in November, are both characterized by extended periods of overcast and light drizzle and by brief periods of intense precipitation.

Reliable climatic data for the Hesperian Meseta are sparse. At best only five years of data have been recorded for "stations" within or near the study area and these records are sporadic and incomplete, since in all but one instance only precipitation has been measured. The available data from these locations are recorded in Table 2 for comparative purposes. Consequently it is necessary to rely on climatic records assembled from outside the immediate area. According to Lautensach's regional delimitation (1964: Map 14), the city of Soria (data from British Meteorological Office, 1958) lies at the northern margin of the Hesperian Meseta while Molina de Aragón (data from Boletín Mensual

TABLE 2

CLIMATIC DATA FROM LOCAL STATIONS
(SHORT RECORDS)

Mean Monthly Precipitation
(in millimeters)

Station	J	F	M	A	M	J	J	A	S	O	N	D	Total
Atienza, 1150 m (3-year mean)	61.7	92.1	37.5	45.3	49.0	68.6	16.5	5.4	50.4	59.6	71.1	50.0	607.2
Cogolludo, 900 m (2-year mean)	54.3	73.0	20.0	--	79.5	25.5	--	25.0	6.0	204.5	74.7	65.1	627.6
Jadraque, 850 m (4 to 10 year means)	60.2	62.9	54.6	50.6	47.1	58.4	15.7	11.2	44.9	70.2	82.7	74.0	632.5
Mirabueno, 1065 m (4-year mean)	61.9	77.9	43.0	76.3	25.4	80.6	19.1	7.6	43.4	65.0	82.8	47.5	630.5
Sigüenza, 1020 m (4-year mean)	72.5	119.0	36.4	85.7	34.4	63.4	21.6	4.4	32.9	67.6	86.6	42.3	666.8
Valdelcubo, 1015 m (4 to 9 year means)	34.0	31.6	22.4	22.3	31.8	45.1	22.1	23.0	35.4	34.9	51.5	48.5	402.6

Climatológico) is situated near the southeastern border. The records of these stations are similar and therefore presumably representative of the general 1000 to 1100 m elevation range, if not of the Hesperian Meseta as a whole. The following discussion draws upon the Soria data because of the substantially longer period on which it is based (Fig. 3).

The mean daily January temperature of 2.5°C (36.5°F) is defined by mean daily maxima and minima of 7°C (44°F) and -2°C (29°F) respectively; the absolute minimum recorded for the 30 year period is -17°C (2°F) while the mean of the absolute minimum temperatures is -10°C (14°F). Statistically, January is the coldest month of the year but temperature parameters for this month are not significantly different from those of December so that, in effect, there are two months of the year when extreme minimal temperatures prevail. This situation is not unlike that which exists during the high-temperature season. Although August is the warmest month of the year--mean daily temperature, 20°C (68°F)--the July parameters are not significantly different so that either of these two months is assured high temperatures. The mean daily maximum temperature in August is 28°C (82°F), while the mean of maximum August temperature reaches 34°C (94°F). The absolute maximum recorded for the observation period is 41.1°C (106°F). The annual march of mean temperatures and temperature extremes is shown for Soria and Molina de Aragón in Figure 3. The mean annual temperature range of ±20°C experienced in the Hesperian Meseta is exceeded in Spain only by that of the eastern Pyrenees and the La Mancha region of New Castile (see Lautensach, 1960:Map 3).

Marked seasonal differences characterize the precipitation regime of this area of central Spain. The extreme aridity of the megathermal summer months is preceded and followed by primary and secondary precipitation maxima respectively (Fig. 4). May is the wettest month of the year, receiving more than 10% (66 mm) of the mean annual precipitation (569 mm); August is the driest month with only 23 mm of precipitation. The second precipitation maximum generally occurs in November (or late October); the city of Soria records 56 mm for this maximum. Measurable precipitation of more than 1 mm is recorded on an average of 111 days of the year. The frequent and abundant autumn dew cover presumably is not included in this total.

A major portion of precipitation occurring during the humid seasons, particularly in the autumn, falls as a prolonged light drizzle during extended periods of overcast or cloudiness. By contrast, the infrequent summer showers characteristically come as brief but intensive downpours. At Molina de Aragón, for example, 56% of the average total August precipitation occurs within a 24-hour period. Personal experience would indicate that in many cases all of the 24-hour total may relate to a single shower of short duration. Climatic records do not reveal this aspect but one study has related this summer precipitation characteristic (Hessinger, 1949). On approximately 10 days of the year precipitation

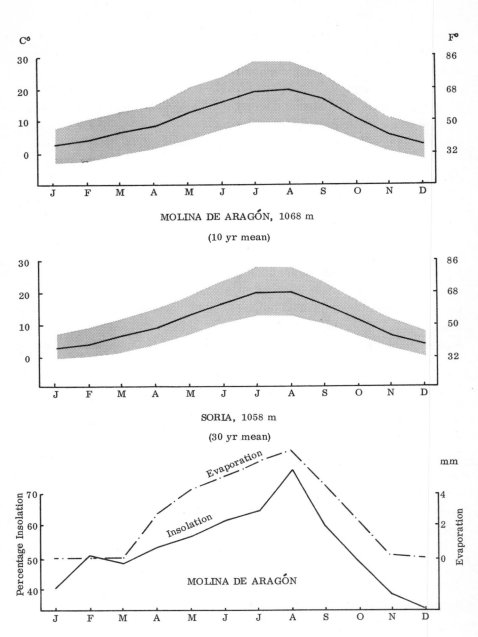

Fig. 3.--Annual temperature curves for Soria and Molina de Aragón. Temperature data are from Boletín Mensual Climatológico (1957-1966) and British Meteorological Office (1958); shaded areas indicate mean daily temperature ranges based on mean daily maxima and minima. Mean insolation and evaporation data are from Landsberg et al. (1965); amount of insolation is stated in mean percentage recorded during sun-up hours.

27

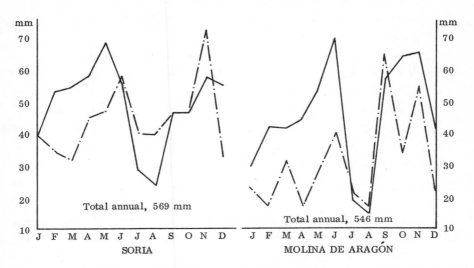

Fig. 4.--Annual precipitation curves for Soria and Molina de Aragón. Data are from Boletín Mensual Climatológico (1957-1966) and British Meteorological Office (1958). Solid lines indicate mean monthly precipitation based on 30 and 10 year records for Soria and Molina de Aragón respectively, and the broken lines show absolute maximum precipitation during a 24-hour period, based upon 27 years and 10 years of records respectively.

at Molina de Aragón falls in the form of snow. Duration of snow cover is short, rarely more than five days (Lautensach, 1964:Map 28b).

According to the data for Soria and Molina de Aragón, the Hesperian Meseta falls within the $C_1 B'_1 sb_4$ climatic type of the Thornthwaite classification, which is a dry-subhumid, mesothermal climate with a moderate winter water surplus and a temperature-efficiency regime normal to the fourth mesothermal (48%-52% summer concentration of potential evapotranspiration). By the Köppen system, this is an area of Csb_3 climate, a summer-dry mesothermal situation with the coldest month between $0°-6°C$ ($32°-42°F$) (see also López Gómez, 1959).

Vegetation

Scrub and woodland are commonly considered as the prevalent plant association of the Hesperian Meseta. It is more accurate, however, to refer to this area as a Mediterranean scrubland since, by virtually any measure, this is the most widespread and characteristic vegetation association. Areas of dense woodland are restricted to hollows that offer protection from the crippling wind, or to areas too impoverished to sustain the minimal needs of pastoralism.[1] This scrubland condition contributes to the increasing awareness of what Houston (1964, 207) describes as "the desolate monotony of the Peninsula's vast, empty spaces."

Spontaneous woodland of the Hesperian Meseta is comprised of pine and oak species, intermixed with juniper, beech, birch or ash, as well as shrub species that reflect local (edaphic) and altitudinal (climatic) factors. As in the case of the scrubland vegetation, each arboreal family retains xerophilous characteristics appropriate to the hydrothermal regime of the area. The following discussion is concerned first with the arboreal species of the study area and then the non-arboreal scrub or matorral community. The spontaneous species considered are predominant in the study area and characteristic of the Hesperian Meseta as a whole. Their distribution shown in Figure 5 is based upon the 1:400, 000 forest map of Spain (Ceballos, 1966).

Extensive stands of Mediterranean black pine (Pinus pinaster, ssp. mesogeensis) are still found as a spontaneous cover colonizing the sandy, silicate slopes of the Buntsandstein terrain. This hardy species is well adapted to the temperature extremes of the Hesperian Meseta and forms real forests on the Buntsandstein exposed northeast of Sigüenza and again southeast of Alcolea del Pinar. Even though forest crowns suggest luxuriant cover, the stands are rarely dense so that additional plants may generate on the forest floor. Where ground moisture is maintained, small ferns survive.

Small stands of scots pine (Pinus silvestris) and lodgepole pine (Pinus laricio, ssp. pyrenaica) have been replanted throughout the Hesperian Meseta at general elevations between 1000 m and 1800 m. Scots pine is lime tolerant but maintains a preference for silicate soils, as shown by its repopulation on Buntsandstein sands east of the headwaters of the Rio Jalón, and west of Barcones. The lodgepole, on the other hand, enjoys a wide tolerance range and consequently has been replanted on shaley Keuper midslopes of the Sierra de Ministra and other parameras throughout the upper Jalón valley, as well as on the upland limestone surfaces themselves. In spite of protection from grazing in these areas, regeneration has been slow in the face of moisture deficits and present growths commonly do not exceed 3 m in height.

Spontaneous broadleaf stands are less dense and individual trees are shorter than in the conifer woodlands. The forest crown is never completely closed so that these woodlands are grazed to utilize the sparse grasses that generate on the forest floor (Fig. 6). Turkey or blanket oak (Quercus pyrenaica=Q. toza) and prickly oak (Quercus lusitanica, ssp. faginea) are the predominant arboreal species comprising the deciduous broadleaf forests. Both are hardy continental types that propagate well between 600 m and 1200 m; contrasting parent material preferences explain local distributions. Turkey oak is not

---

[1]An obvious exception is the areas of repopulated national forest.

29

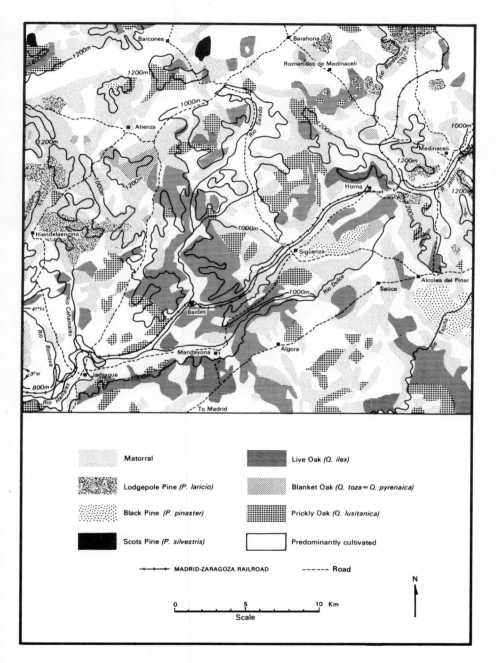

Fig. 5.--Distribution of dominant vegetation types in the study area (after Ceballos, 1966).

Fig. 6.--Páramo vegetation. Incomplete woodland cover is provided by <u>Q. ilex</u>. Limestone detritus intermixed with a thin <u>terra rossa</u> soil mantle, and micro-lapiés features etched into the Pontian limestone bedrock are evident in the foreground.

lime tolerant and spontaneous stands in the study area are restricted to Paleozoic shales and schists south of Atienza and to the northeast flank of the Sigüenza anticline. Prickly oak shows a preference for calcareous material; small stands have developed on the Liassic limestone surfaces above Romanillos de Medinaceli, on the Silurian shale core of the Riba de Santiuste anticline, and locally on the Pontian limestone páramo.

The dominant arboreal species of the Hesperian Meseta and of the Mediterranean Basin in general is the ubiquitous live or holm oak, <u>encina</u> (<u>Quercus ilex</u>). This broadleaf evergreen is exceedingly hardy and adapted to a wide range of conditions and associations. Within the study area, it is found in association with <u>Pinus</u> and other <u>Quercus</u>

species, and the matorral. Most frequently it occurs on limestone material where it is dominant within a community. Live oak and lodgepole or scots pine occur with scrub forms of juniper (Juniperus thurifera), an extensive stand of which covers much of the 1200 m páramo north of Maranchón. Beech (Fagus silvatica), birch (Betula verrucosa) and ash (Fraxinus angustifolia) also occasionally appear in these higher upland situations. Stands of poplar, chopas (Populus tremula), fringe the lowland streams to complete the arboreal picture.

Most of the spontaneous cover of the Hesperian Meseta is comprised of the non-arboreal matorral, a sclerophyllous shrub association variously called maquis, macchia or garrigue in the Mediterranean lands. It has a close ecological relation with the chaparral of the southwestern United States. Four families are generally recognized within the matorral communities: Ericaceae, Cistaceae, Labiatae and Leguminosae. In the Hesperian Meseta, ericaceous heath or brezales includes arborescent (Erica arborea) and broom or escoba (Erica scoparia) species together with acidophyllous plants, such as the ling (Calluna vulgaris), and purple bell-heather (Erica cinerea). Additional broom species recognized in the eastern foothill section of the Cordillera Central include Sacothamnus scopanus, S. patens and Cytisus multiflorus. The cistaceous or rockrose family components of heath, called garabes or jarales matorrales by the Spanish, include Cistus laurifolius, C. Ladaniferus and C. monspeliensis, all of which exhibit mountainous preferences on silicate material.

The pleasant aroma which pervades the Hesperian Meseta derives from the dead nettle family (Labiatae). These matorral constituents include lavender (Lavandula spp.), marjoram (Thymus mastichina), rosemary (Rosmarinus officinalis), sage (Artemesia spp.), and thyme (Thymus vulgaris). Additional Labiatae include: Gensita scorpius, Rhamnus lycioides, Phillyrea angustifolia, and Cistus albidus. Broadleaf shrubs and thorns such as Acacia spp., Mimosa spp., and Cassia spp., gorse and lark (Ulex spp.), as well as hawthorn (Calvcotome spinosa), comprise the Leguminosae of the study area.

The matorral community occurs throughout the dry sector of the Iberian Peninsula, varying in composition and density with parent material, local moisture factors, and degree of human interference.[2] It is held by some to be a climax form in parts of Spain (see Ceballos, 1966:49f; Eyre, 1963:126); these occurrences are limited to areas where because of altitude, soil salinity, or other factors, forest colonization is impeded or impossible. In swamp and salt marsh situations, such as gypsiferous springs and seeps,

---

[2]Hartmut Ern's (1966) recent examination of vertical zonation in the Sierra Ayllón suggests that the matorral may be characterized by numerous additional sclerophyllous species; however, direct analogies would be suspect since he has dealt with an area that has undergone considerably less human interference, and that presumably is also more representative of the intermixing of Atlantic and Mediterranean species.

this climax form may prevail; however, the principal growth is found in the high sierras, above the tree line, where wind, temperatures, soil, and snow conditions are too severe for juniper or broom species.

The climax vegetation of the Meseta is by no means firmly established or generally agreed upon, but the matorral that dominates most of the expanse is held generally to be a spontaneous form (see Solé, 1952:260-68). Hopfner (1954) has pointed out that wholesale deforestation of Old Castile dates from historical times and that major portions of this were accomplished since the Reconquista (11th-12th centuries). Repeated burning and grazing by pastoralists as well as local deforestation since medieval times have encroached upon and destroyed the natural woodlands of the Meseta, while continuous use of these lands has insured the regeneration of the regressive matorral. According to Ceballos (1966), where burning has persisted, Cistus species assume regenerative dominance while Erica species dominate those areas cleared by deforestation.

The principal types of regression and their associations are commonly identified throughout the Mediterranean Basin by their chief plant and at least nine such types can be recognized (Houston, 1964:87). Ceballos (1966) identifies four such phases for dry Spain ranging from a dominance of arboreal species to a dominance of shrubs and heath-type plants. These types, as well as their transitional stages, reflect the interplay of natural conditions of regeneration and cultural changes. As a consequence of the complex nature of the types of degraded vegetation, their peculiar associations and stage of regression represented, the definition of a matorral becomes a little arbitrary. For this reason, it is most appropriate, particularly within this geomorphic context, simply to consider matorral as a xerophytic shrub association that provides an incomplete vegetative cover for most of the uncultivated lands of the Hesperian Meseta.

Nowhere in the Hesperian Meseta does the vegetation cover provide a complete ground cover (Fig. 7). Within the woodland areas the moderate or low tree density and open crowns provide limited protection of the forest floor, although in general terms the percentage of ground cover is greater in Pinus woodlands than in the Quercus. In spite of sun penetration, climatic and edaphic aridity deter luxuriant undergrowth. Forest litter is sparse and usually limited to thin localized accumulations of conifer or deciduous oak leaves. Consequently in most wooded areas the ground surface is susceptible to a certain degree of overland flow and erosion after the raindrop impact has been absorbed at the forest floor. In the case of Quercus woodlands that populate limestone surfaces, overland flow is short lived in the face of absorption and storage capacity factors.

Erosion is potentially more severe in the matorral. The interspaces between shrubs are void of vegetation and the soil is friable or loose as a result of the incessant disturbances of pastoralism. Such surfaces are readily affected by raindrop erosion

Fig. 7.--Woodland vegetation. The open forest crown is provided by black pine (P. pinaster) through which the páramo surfaces can be seen at the horizon.

while diffuse runoff may assume the form of surges, interrupted and temporarily impeded by the shrubs. The limited ability of the matorral to inhibit erosion is apparent from the exposure of upper portions of root systems.

Shallow arboreal and non-arboreal root penetration provides only limited cohesiveness in the active horizons of the soil solum. Consequently the ground cover is particularly susceptible to mass movements. Apart from the ubiquitous and slower dry-soil forms, such as creep, earthflows become significant where denudation is unimpeded by deeply anchored root systems, particularly in the Keuper shales. Full assessment of the impact of cultural factors on the Holocene geomorphic activity of this area must await more detailed sedimentological studies.

Soil Mantle

Detailed investigation of the soil mantle of the Hesperian Meseta has yet to be carried out and soil mapping is confined to small-scale, generalized maps of the entire Iberian Peninsula (see del Villar, 1937:1,500,000; Kubiena, 1956:1:4,000,000; Guerra, 1968:1:1,000,000). These studies show dry, calcareous terra rossa soils and deep rotlehms areally dominating the core of the Hesperian Meseta, with lithosols on the more extensive outcrops of Buntsandstein sands. The Hesperian margins of the Somosierra and Serranía de Cuenca have braunlehm sediments and humid or subhumid rendzinas respectively. The Duero, Ebro and Tajo basin-heads adjacent to the Hesperian Meseta are mantled by xerorendzinas and deep, earthy, carbonate-rich terra fuscas. Almost all of these soil types are relict forms that no longer form afresh in the contemporary environment. Soil formation is widely thought to be quasi-dormant throughout the Mediterranean pedogenic region to which these varieties belong.[3]

Summer drought and winter cold effectively inhibit pedogenesis although chemical weathering is not entirely absent. Soils that have developed on post-Pleistocene alluvia under woodland (rendzina, ranker or braunerde types) do not exhibit the deep chemical weathering of the relict soils nor, in the face of widespread deforestation and erosion in historical times, do the relict soil profiles appear to form anew. In general, biochemical weathering is also curtailed as shown by the impoverished nutrient cycles of the Hesperian Meseta and the predominance of a xerophytic-sclerophyllic vegetation cover. Micro-organic activity is limited in variety and intensity.

The actual contemporary soil mosaic is, of course, complicated and at this point only generalizations can be made. The following brief résumé of soils derives primarily from Kubiena (1953, 1956) and del Villar (1937), as well as from personal field observations. The remarks pertain to the study area specifically although they also may be presumed to be relevant for the Hesperian Meseta as a whole.

Non-calcareous soils developed on silicate rocks are primarily of the ranker type. Rankers occur on the Paleozoic quartzites, shales and micaceous schists of the Riba de Santiuste anticline and among the Somosierra foothills immediately west of the study area

---

[3]Various authors have demonstrated the relict nature of the terra rossa-terra fusca and allied rotlehm-braunlehm red soils (see Klinge, 1958). Pedogenesis of the latter types has been related to Mio-Pliocene times in Central Spain (Kubiena, 1954), and to Pleistocene times in the Costa Brava area (Butzer, 1964a:42-49) and on Mallorca (Butzer and Cuerda, 1962). By implication, development of the former types on calcareous parent material was contemporaneous (see also del Villar, 1937:199ff). Meridional braunerde on silicate rocks (Butzer, 1964a) and rendzinas or calc-braunerdes on calcareous material (Kubiena, 1954; Butzer, 1963b; Brunnacker and Ložek, 1969) represent post-Pleistocene climax forms.

(see del Villar, 1937:115ff and map). Weakly developed ranker soils also occur locally on Buntsandstein sandstone outcrops. These soil mantles consist of thin, nutrient-deficient A-horizons in stoney or granular admixtures, depending upon the parent material, directly overlying the C-horizon. Ranker enclaves are particularly patchy in occurrence, due primarily to their non-cohesiveness and erodibility, and may appropriately be referred to as calvero-ranker soils (see Kubiena, 1953:274). Intermediate and footslope zones of the silicate bedrock areas comprise lithosols and older colluvial materials. In the western margins of the Hesperian Meseta, such colluvial materials are likely to include sediments derived from braunlehms and meridional braunerdes, soil types that once formed on the southern slopes of the Sierra de Guadarrama. These occur as truncated relics today (Kubiena, 1953:293ff).

Rendzina and terra rossa types are the calcareous soils most characteristic of the upland surfaces. Rendzina soils have developed thin to moderately deep profiles on most of the Mesozoic limestone of the study area. The humus-deficient mull or mull-like, loose or powdery A-horizons are seldom deep and pass directly to C-horizons, although Ca-horizons may intervene where soil development is sufficiently thick on hard parent material. These soils most closely resemble xerorendzinas (Kubiena, 1953:189f). Higher clay contents produce mull rendzina varieties, an example of which is described by Alvira and Guerra (1951) from the southern margin of the Hesperian Meseta near Molina de Aragón. Xerorendzina catena transitions to the sierosem undoubtedly occur within the Hesperian Meseta, particularly in relation to gypsiferous parent material of Tertiary or Triassic age.

The widespread development of rendzina soils on Mesozoic limestones is noteworthy. Both Kubiena (1953) and del Villar (1937) show relict terra rossa soils as the predominant mantle of the study area but during the course of this field work the typical decalcified, brick-red, ferric oxide rich, surficial mantles of the terrae calxis group were not encountered on these limestones. Del Villar (1937, 260) has suggested that mull-like rendzina humus under woodland or dense undergrowth often masks older terra rossa soils in the limestone country of dry Spain but no situations of this sort were observed in the study area. In fact, where excavated, moderately deep (ca. 50 cm), mature AC rendzina profiles were uncovered developed on the upland surfaces of Mesozoic limestone. These soils are generally held to be the climax form of the Spanish Meseta (see Klinge, 1958:60, and previous footnote references) but their degree of development and disposition on surfaces adjacent to and higher than the extensive relict terra rossa mantles of the Pontian limestone suggest that these profiles are relatively ancient.[4] The precise age of

---

[4]Klinge (1958, 90f; 106f) described rendzina horizons on terra rossa soils in Triassic and Pontian limestone country of southern Spain in situations not unlike that of the

these soils remains, however, an open question.

Rendzina-type soils also occur in relation to some Quaternary materials. Travertine terrace deposits of the Alto Henares system now commonly have grayish, loosely structured, carbonate-rich, A-horizons of this type in association with xerophytic matorral vegetation. The organic component is generally increased as a result of pastoralism. These horizons occasionally occur on more recent floodplain deposits rich in carbonates and, where clay enrichment has been sufficient, the development of a crumb structure and mull humus with a lower carbonate content provides mull rendzinas.

Relict terra rossa soils comprise the Pontian limestone mantle. Locally, soil profiles may exceed 2 m in depth, but more characteristically, humus-poor, decalcified, red A-horizons of 5 to 10 cm overlie deep, vivid red, (B)-horizons rich in ferric hydroxide. The latter horizons contrast strikingly with the white C-horizons in chalky limestone. The intensive red of these terra rossas provides the colorful carpet for much of the Castilian wheatlands, while terra rossa-derived sediments contribute in-fill to incipient depressions in the limestone surfaces as well as to the páramo slopes.

Lithosols and soil-derived sediments mantle the slopes below the páramo cuestas. In the Mesozoic geologic province, terra rossa sediments form recalcified colluvial mantles locally up to 2 m thick. Humification is very limited because of slope erosion and open vegetation cover. Similar colluvial mantles cover the midslopes of Tertiary conglomerates, silts and marls. The humified, loose-structured, sandy rendzina-like A-horizons generally have higher carbonate content because of the marl bedrock.

Intrazonal soils occupy the valley bottomlands which include local tracts of alkaline gleys of the solonchak type. These are particularly prominent in the salado-saladillo (salt river, or brook) floodplains developed amid Keuper shales. Light to dark gray, chalky humus is characteristic with superficial efflorescence of salts that undergo wet-dry season vertical migrations. As del Villar (1937, 353) has emphasized, these saline patches of interior Spain are not the consequence of playa-type development but reflect local lithologic conditions. The solonchaks are subject to seasonal inundation or surface desalinization and the halophytic vegetation may support small herds of cattle. Colluvial materials interfinger with floodplain deposits at the campiña margins. The relationship of the several soil types discussed above is shown schematically in Figure 8. It must be emphasized that this is a generalized representation based upon a scattering of field observations

---

Hesperian Meseta. If rendzina pedogenesis post-dates the terra rossas, then the latter soils may also have developed on Mesozoic limestone in the study area, in which case they were almost completely stripped, perhaps in association with the Pliocene planation cycle discussed in Chapter V. In any case, the superposition of calcareous AC rendzina profiles upon decalcified A(B)C terra rossa soils should pose interesting questions for the pedologist.

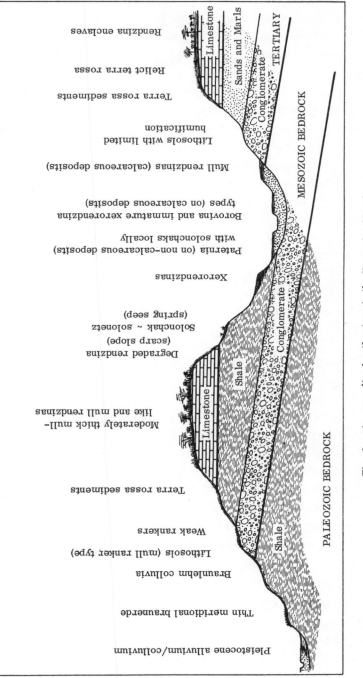

Fig. 8.--A generalized soil catena in the Hesperian Meseta

and upon the available literature. Thorough study of soils in the Hesperian Meseta undoubtedly will refine the idealized catena.

With few exceptions, the soil mantle is thin and widely susceptible to erosion, so that soil profiles commonly are truncated. The ineffectiveness of this mantle and the vegetation cover to inhibit fluvial erosion subjects the land form to modeling in excess of what might be anticipated under the present semiarid climate. It has been estimated that all cultivated slopes in central Spain of more than 3% are subject to erosion. Erosion on slopes as gentle as 5% is sufficient to destroy organic horizons and in some cases even the remaining horizons of the profile (Bennett, 1960). Historic destruction of woodland cover in central Spain has also disrupted the vegetation-soil symbiosis, so that eroded humus horizons are common phenomena accompanying the forest-matorral regression. As a result, ultimate alteration of the soil structure, impoverishment of the nutrient cycle, and reduction of clay-humus content place increased emphasis on the role of the mineral parent material, rather than the organic component, in providing a binding or cohesive fabric to resist erosion. In general, limited clay content, low plasticity of sandy or silica-poor soils, or low porosity variously contribute to erodibility. Climatically-curtailed pedogenesis, the essentially open vegetation cover, and traditional land-use practices are the major offenders contributing to the ubiquitous erosion of the Meseta regions.

Contemporary Geomorphic Processes

Weathering

There are no data on rates of weathering or rates of denudation for the Hesperian Meseta. The remarks offered in the subsequent sections draw upon personal observation and relevant literature, and implement inferences consistent with the geomorphic constraints outlined in the previous sections of this chapter. Of interest here are the types and extent of weathering processes operative under present conditions. Solution, hydration, oxidation and hydrolysis all may be recognized as playing varying roles in the chemical breakdown. The following assessment of chemical weathering in the Hesperian Meseta is based upon minerals in solution as determined for several springs throughout the study area.[5]

Chemical weathering

The major mineral constituents of stream waters are magnesium, chlorine and calcium products in solution; traces of sodium chloride and sulphur trioxide also occur

---

[5]Data are from the Mapa Geológico de España Explicación: Barahona, 434; Sigüenza, 461; Maranchón, 462; Jadraque, 486; Ledanca, 487.

frequently. These minerals are derived for the most part from attack of the limestone caps by carbonation. Since carbon dioxide concentration is inherently high (>40%) in the limestones of the study area (de la Concha, 1963:35), rain and ground water produce ample quantities of carbonic acid for solution. Carbonation undoubtedly is particularly effective since precipitation maxima coincide with the cold season when CO_2 concentrations normally are highest. Water-storage capacity of the meseta limestones is high, as shown by the perennial nature of the springs that issue from the limestone basal contact plane, so that this type of solution may occur throughout the year. Solution by hydrochloric acid also plays a role in chemical weathering. The chlorine comes from the simple solution of salts, primarily halite from the Keuper strata. Intense hydrolysis, which requires leaching by fresh water, is inhibited by the scanty annual precipitation and aridity during the warm season. However, the higher temperatures and increased daily temperature ranges of this season, in combination with additional H ions released by carbonation, furnish the ingredients necessary to promote hydrolytic reactions, if only to a limited degree (see Keller, 1957:29f). Oxidation occurs in the alteration of iron and aluminum traces in the various limestones to form sesquioxides, Al_2O_3 and Fe_2O_3 respectively, and sulphuric acid formed as a result of oxidation of sulphur dioxide in some limestones acts as an additional solution agent for chemical weathering.

In overview, it appears that in spite of the semi-aridity of the Hesperian Meseta and the coincidence of the dry and warm seasons, each of the general forms of chemical weathering is presently operative. However, these are effective to different degrees and at rates that may vary seasonally. Carbonation is presumed to be the predominant process, most active during the cooler part of the year, particularly during the two wet seasons. Hydrolysis and oxidation, on the other hand, are of lesser overall importance and are less variable in their seasonal rates of weathering. Aside from these relative assessments no absolute statements concerning the degree of weathering can be offered. However, in general, the low clay content of most soils and the limited quantities of minerals in solution indicate that chemical weathering in the Hesperian Meseta is not substantial under present conditions.

Physical weathering

Limestone detritus commonly exhibits sharp, angular edges as a consequence of physical weathering, although the mechanics of the process involved are not clear. It is questionable whether frost-shattering occurs since cold-season temperature amplitudes are insufficient for dramatic water-ice crystal expansion; however, daily winter freeze/ thaws, in combination with heat absorption by exposed bedrock surfaces, may engender thermoclastic weathering. Frost-wedging in joints of the limestones presumably has aided in the breakdown and collapse of massive limestone blocks undermined by mass wasting of the underlying saturated shales. In any case, frost shattering must have been involved in producing the detrital limestone fragments that litter most midslopes of the Triassic terrain. It is also reasonable to assume that physical hydration plays a role in the mechanical weathering of those Triassic and Tertiary sediments with salts and high clay contents. Some evidence of this was observed with respect to the Keuper shales. Within a period of 18 months, spanning two winters, a macadam road surface in the study area was rendered almost impassable as a consequence of heaving produced by physical hydration and frost action. Data are not available but the moderate climate and modest regolith mantle suggest that physical weathering plays a minor role under present conditions.

Erosion

Mass wasting

Earthflows, slumps and debris slides are the most common and most important forms of mass wasting in the study area. Such slippage has been observed only in the Mesozoic bedrock terrain where its role in undermining the upland limestone caps has already been noted. Overly steep Keuper midslopes become unstable when saturated. This commonly occurs directly below the limestone-shale contact plane where wet conditions prevail because of the moisture storage capacity of the limestone. Steep slope segments, shallow matorral root systems, and thin soil mantles further contribute to earth flowage in these upper segments of the overall slope profile (see discussion by Bailey and Rice, 1969). A particularly large slippage scar can be seen below the páramo crest at 1000 m south of Sigüenza. Collapse of saturated Keuper slopes is very common on a smaller scale along road cuts. The stabilizing effect of vegetation is apparent when older, vegetated road cuts are compared to recently excavated barren ones, which are prone to slope failure and require frequent regrading. In time, however, equilibrium seems to be achieved since older road banks with vegetation maintain a protracted stability that is disturbed only during particularly wet cycles.

Evidence for soil creep is less widespread and limited to rare situations of deformed root systems and to the personally observed downslope movement of detritus in response to gravity. Thornes (1967, 89f) cited similar circumstances in his study in northern Soria but was unable to assess the relative importance of this form of mass wasting.[6]

---

[6] Maxwell (1967, 181) noted the operation of creep in chaparral vegetated watersheds in the San Gabriel Mountains of southern California, but again no data were included. An additional surface form of interest described in this study (Maxwell, 1967:175) is a "debris chute," which is a linear depression caused by downslope rolling and sliding of surface

Frost action may be assumed to play some part in the downslope movement of regolith since frost activity appears to be involved in the physical weathering of some bedrock materials. Thornes (1967, 90f), assuming the frost movement of soils to be a function of grain size, water availability, freezing rate and pressure, and additional variables, postulated that weathered Cretaceous (Cenomanian) limestone, and Tertiary marls (Miocene) and limestone (Pontian) in the Soria area are susceptible to frost movement, but data are still lacking.

Fluvial erosion

Subaerial erosion is seasonal and highly variable because of the marked seasonality of the precipitation regime. Moisture utilization demands further limit water surpluses so that protracted surface runoff normally occurs during the spring. This is shown in Figure 9 which graphically depicts the water balance of the Hesperian Meseta as determined by the Thornthwaite (1948) method. Actual moisture surplus exists for a third of the year or less. In the case of Soria, a surplus is maintained for the January through April period (93 mm), whereas at Molina de Aragón, January through March are surplus months (67 mm). Water balance deficits of 166 mm at Soria and 168 mm at Molina de Aragón are sustained from July through September. Surplus in terms of stream discharge attains highest levels during the late winter and early spring months, February through April, as indicated by stream flow gauge readings for approximately 3600 sq km of the upper Rio Tajo drainage. These graphs show that runoff and stream discharge attain maximum levels during early spring and late fall when, according to the Tajo discharge curve, a substantial portion of the precipitation not only recharges ground storage but also contributes directly to surface runoff. The data suggest, therefore, that the major work of fluvial erosion is accomplished during these periods. However, the intensity of the infrequent dry season downpours also accomplishes impressive amounts of denudation. One such storm (September 2, 1967) of less than one hour duration caused widespread flooding and erosion: extensive rill cutting and gullying altered low-order drainage patterns; rills up to 65 cm developed in gentle 3° slopes, with discharge competence sufficient to transport cobbles of 15 cm in diameter; ephemeral depositional features in rills and channels indicated large amounts of suspended sediment. It is very probable that such torrential downpours toward the end of the dry season account for a major part of erosional activity.

Denudation under present conditions involves sheet wash, linear erosion and gully head extension. Sheet erosion, best observed on the dip slope of Muschelkalk limestone,

---

debris. Similar features, sometimes choked with detritus, can be observed in the Hesperian Meseta area. But since these rills carry channel flow when there is surface runoff, it is not entirely clear if their origin can be tied to mechanical rather than fluvial erosion.

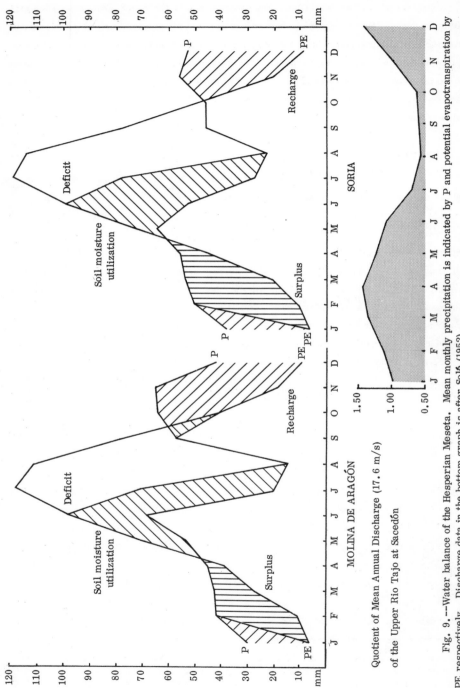

Fig. 9.--Water balance of the Hesperian Meseta. Mean monthly precipitation is indicated by P and potential evapotranspiration by PE respectively. Discharge data in the bottom graph is after Solé (1952).

assumes the form of surges temporarily dammed by exposed _matorral_ root systems and surface litter. This type of erosion occurs during the dry season downpours; sediment yields have not been determined but other researchers (Leopold et al., 1966) have shown sheet erosion to be the major contributor of sediment in an analogous environment in New Mexico. Linear erosion comprises rill cutting and gully enlargement in the weaker shales and marls of the Triassic and Tertiary sequences. Incision is active on unprotected slopes throughout the wet seasons. Even low angle slopes undergo rill cutting, particularly when cultivated since furrows commonly run perpendicular to slope contours. Rill networks are ephemeral, disappearing and redeveloping with successive rains, but nonetheless effective in transporting fine sediments downslope as shown by miniature alluvial fans developed locally within the rill channels. This form of linear erosion is perhaps the major agent of soil erosion in the Hesperian Meseta.

Systematic observations on stream channel characteristics and sediment loads are difficult to make for the study area because of widespread modification of drainage channels for irrigation purposes. Nonetheless, some conclusions regarding stream transport can be reached on the basis of personal observation. Throughout most of the year perennial streams are ostensibly free of suspended load. Transport must therefore be confined to material in solution and to a very limited transfer of bed load. During periods of intense or extended precipitation the levels of suspended sediment increase dramatically, in response to both denudation and increased stream capacity. Stream-bed deposits in general reflect the predominant character of local geologic facies. Where coarse materials from Triassic and Tertiary conglomerates are immediately available, stream beds contain a preponderance of gravels; with fine-grained bedrock, sands, silts and muds are characteristic with little or no coarse bed load.[7] Knickpoints occur in the smoothly concave longitudinal profiles of the Henares, Salado and Dulce rivers where they traverse the Mesozoic-Tertiary geologic contact. Upstream from the break in profile, stream-bed deposits conform closely to the above generalization. Downstream, however, broader meandering floodplains have developed commonly with fluvial gravel deposits irrespective of local bedrock facies. These are older deposits unrelated to present stream alluviation as shown by their lateral conformity with buried gravels exposed along present stream banks. Consequently present stream energy is primarily expended in the reworking of older alluvia.

---

[7] Within the study area there are no perennial streams which have not cut through the upland limestones. Dissolved load appears to be dominant in channel beds near these caps, with some limestone pebbles on the stream bed. If the local limestones are marly, pebbles are largely replaced by limey muds carried in suspension.

In overview, it seems that in spite of limitations imposed by sparse precipitation and mesothermal temperatures, significant morphologic modeling obtains under present-day conditions. Empirical data are not available and quantitative estimates of degradation can not be made. Man has been a particularly effective geomorphic agent in this environment through his disruption of the vegetation cover so that assessment of denudation rates must account for this variable.

CHAPTER IV

LAND-SURFACE FORM

Introduction

Land-surface form of the Hesperian Meseta is dominated by the flat, upland plain characteristic of all tablelands. This morphologic element of the Mesozoic plateau is common to both the Castilian tablelands and the Hesperian Meseta. Dissection of the limestone-capped plateaux and the enlargement of drainage basins hollowed out of the weaker underlying materials have produced areal variations in land-surface form. The array of the land-surface form encountered within the tableland and the distinctive geomorphic elements of the study area are the focus of this chapter.

Land-Surface Form

The variety of surface morphology is depicted in Figure 10 which encompasses 2525 sq km of the Hesperian Meseta. Data on which the map is based were compiled from 1:50,000 topographic maps with 20 m contour intervals. Local relief, amount of gentle slope, and occurrence of gentle slope in the generalized slope profile were determined independently for unit areas of one square kilometer.[1] Class limits within a category (terrain property) may overlap so that the definition of a land-surface form type derives from the composite of the attributes measured. For example, the amount of gentle slope and profile-type class intervals vary among upland plain, plains with hills, and hills, but local relief is the ultimate limiting variable since class intervals of the first two properties are in part inclusive of one another.

Four types of land-surface form emerge using this system: upland plains or plateaux (páramos); plains with hills, or dissected plateaux (parameras); hills (serranias); and lowlands with hills (campiñas). The parenthetical expressions refer to Spanish terminology that is suggested here as being the appropriate equivalent. The definition of the landform categories is shown in the legend of Figure 10 and a brief discussion of each follows.

---

[1]The procedure used here is based in part on the scheme devised by Hammond (1964).

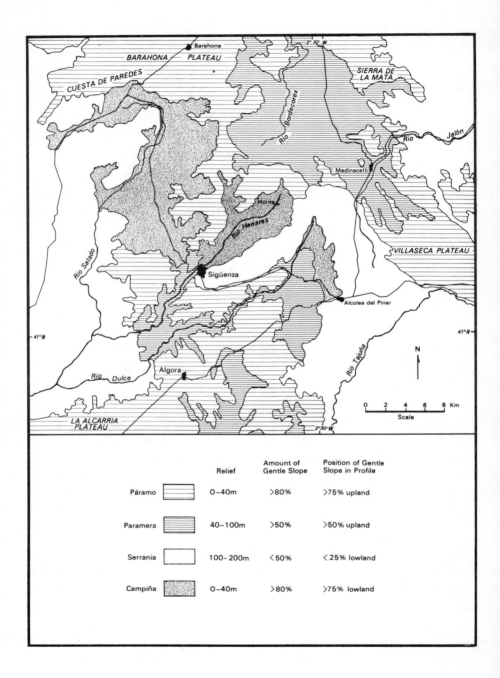

Fig. 10.--Landforms of the study area

Upland Plains (<u>Páramos</u>)

The three expansive upland plains or plateaux that dominate the area mapped are the Hesperian Meseta counterparts of the páramo surfaces prominent throughout the Castilian Mesetas.[2] The Alcarria, the largest, is an almost featureless surface gently beveled to the southwest and extending 120 km to the environs of Madrid. At its greatest breadth, it is approximately 35 km wide. The Rio Henares and its left bank tributaries have incised their courses as much as 200 m into this Pontian limestone surface to effect the angular or subangular crest of the northwest face. Downcutting by the Rio Tajuña has been less aggressive so that the southeast face of the Alcarria páramo is convex-concave in profile.

The Villaseca plateau, located along the eastern margin of the area mapped, is the second of these páramos. Most of the approximately 100 sq km of this surface lies above 1200 m, and several remnants of an older surface exceed 1350 m. The Liassic limestone comprising this páramo is beveled to the north and large segments of the surface are undissected by drainage channels. Dissection of the northwest face by the headwaters of the Rio Jalón, aided by collapse of the underlying weak Keuper shales, has effected subangular crest slopes while the remaining borders break gently into the surrounding lower hills.

Finally, the northward sloping surfaces of the Barahona plateau between 1100 and 1200 m encompass almost 400 sq km planed toward the Rio Duero. Remnants of an older surface above 1200 m occasionally break the horizon, but otherwise areas up to 6 sq km are flat with local relief of less than 20 m. The subangular southwest crest slope, called the Cuesta de Paredes, has resulted from headwater backwearing by the Rio Salado, a major right bank tributary of the upper Rio Henares. Incipient headwater dissection by the Rio Bordecorex effects greater local relief in the central segment of this plateau, but the monotony of the upland is reassumed on the Pontian limestone surfaces of the Sierra de la Mata and Sierra de Muelo páramos. Dissection and retreat of the eastern face has produced the Tertiary hills of the middle Jalón basin.

The páramo surfaces are the distinctive elements of the land-surface form common to both the Hesperian and Castilian Mesetas. It is these features that establish the mesa form that is characteristic of the tableland landscapes. The widespread disposition of the páramos and the disagreement surrounding their development prompt the more detailed discussion of the following chapter.

---

[2]The word <u>páramo</u> as used in Spanish literature refers to desolate, abandoned, flat upland terrain (see de Novo and Chicarro, 1957:24). Its diminutive, <u>paramera</u>, relates to smaller but comparable forms subjected to greater dissection. Some authors, notably Schwenzner (1937) and Houston (1964, 169ff), have used <u>páramo</u> to refer to the Pontian limestone caps. The Spanish usage is preferred here whereby <u>páramo</u> designates a morphologic feature regardless of the age of the materials comprising it.

48

Plains with Hills (Parameras)

Plains with hills occur where mean slope values have increased so that less of the area remains in gentle slope, and where local relief values are significantly greater than in the case of the páramo. These conditions prevail along the dissected margins of the páramo and along the deformed Mesozoic-Tertiary geologic contact, where structural controls have allowed more aggressive fluvial incision of the plateau surface. Locally where the limestone caps have been completely stripped away, the underlying, more erodible Keuper shales, Toarcian limey red sands, or Miocene clays and marls have been remodeled to form more broken topography. Smooth, concave slope forms prevail in these areas along with steeper slope forms and greater local relief than in the case of the páramo surfaces.

Hills (Serranias)

Where the limestone mantle of the tableland has been more completely or even totally destroyed over considerable areas, prolonged dissection and modeling of the underlying materials has produced even greater local relief. Profile types are varied within these provinces but most frequently the gentle slope segments are preserved in the upland portion of the overall slope profile, suggesting at least a local absence of downwearing in the evolution of the present land-surface form. Maximum local relief in these situations approaches 200 m and slope values may exceed 70% (35°). Within the dismantled anticlines near Sigüenza and near Riba de Santiuste, the conglomeratic sandstone hills sustain bizarre features etched into the bedrock cores while in the Tertiary terrain, the exposed marls and clays have been modeled to form rolling interfluves, isolated by smoothly concave interspaces and local badland terrain. Where the dense conglomerate strata are prominent, broadly horizontal benches or platforms occur.

Lowlands with Hills (Campiñas)

In a literal sense, a campiña is a flat, valley bottom of arable land almost always occupied by garden or huerta cultivation. The term is used in a less restricted sense throughout this study to apply to the general expanse of the valley bottom whether flat, gently undulating, or broken by terraces or residuals.[3] All three situations prevail within the Hesperian campiñas but most commonly the valley bottoms are void of dramatic relief or distinctive morphologic features. Pleistocene terraces, considered in subsequent chapters, are only rarely preserved by sharp morphologic definition and the campiña

[3] The single exception to this usage is the campiña terrace referred to in Chapters VI and VII. In that context, the more literal meaning of the term is intended.

floors merge laterally with plateau midslopes by gentle convex changes in slope. In over-all profile, the Hesperian Meseta valleys assume the shape of elongated saucers, or sink-like basins, depressed into the broadly horizontal surface of the plateau upland. The distinctiveness of the campiña landscapes derives not from morphologic features but rather from the intensive agricultural utilization of these areas and the stands of poplar along the banks of the meandering river courses in the valley bottom lands.

The important role of geologic structure in the development of these four land-surface form categories is implicit throughout the foregoing discussion. The effect of par-ticular lithologic units upon the landscape also has been emphasized in Chapter II. Mesozoic and Tertiary limestones comprise the páramo surfaces. Dismantling of this cap has allowed more aggressive dissection and denudation of the underlying clastic bedrock, resulting in the increased local relief and the steeper slope forms of the paramera and serrania morphologic classes. Moderate local relief values of the paramera reflect its disposition at the páramo surface margins that are undergoing destruction. Prolonged flu-vial attack and modeling has effected the more extreme dissection of the serrania landform areas. Enlargement of valley bottoms in clastic materials by higher order drainage chan-nels, in combination with local alluvial and colluvial in-fill, has produced the relatively subdued terrain of the campiña areas.

Further distinctions among the land-surface form types is afforded by the data in Figure 11. The slope values were derived from clinometer measurements in the field, and the idealized interfluve-valley profiles also are based upon field observations. The information summarized in this table augments the class definitions of Figure 10 and allows more meaningful distinctions among the landform categories. Note that even though there is less area in gentle slope in the serrania land-surface form class (see leg-end of Fig. 10), overall mean slope values are less than in the case of the paramera class. This reflects the fact that interfluves are more openly spaced and intervening valley bot-toms are broader than in the case of the paramera terrain. The lithologic aspect of mean slope values is summarized in Figure 12. The values shown here relate closely to com-parable values determined by Thornes (1967, 136) for the Soria area. Inspection of Fig-ure 12 reveals that midslope angles are essentially constant irrespective of the presence of Mesozoic limestone caps, whereas their values vary appreciably in the case of the Ter-tiary bedrock terrain. Presumably both slope retreat and downwearing have been opera-tive in the Hesperian Meseta.[4] Additional features of the land form not disclosed by these analyses are discussed in the subsequent section.

---

[4]Thornes (1967, 130-70) identified four segments of the slope profile and attempted to assess lithologic influence on the variability of these slope angles. After representing these differences with selected field transects, he was able to conclude only that slope modification presently is most active on the clastic materials but essentially inoperative on limestone lithologies.

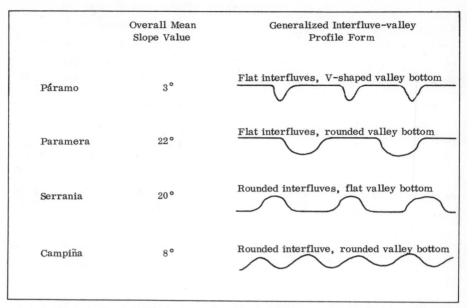

Fig. 11.--Slope profile forms in the Hesperian Meseta. Generalized profiles after a landscape classification scheme proposed by Ollier (1967).

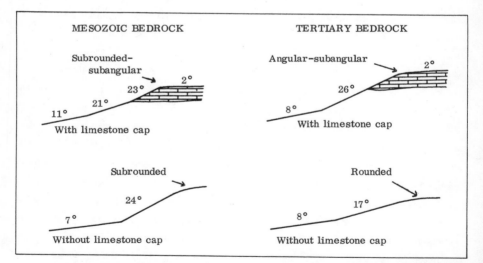

Fig. 12.--Mean slope values in the Hesperian Meseta. Word order defining crest-slope form indicates frequency of occurrence.

Other Landform Aspects

Flat-bottomed Dry Valleys

Flat-bottomed, low order, dry valleys occur throughout the Hesperian Meseta. These features are rarely more than several hundred meters in length or 100 m in breadth. At their margins, abrupt breaks in slope--frequently accentuated by cultivation practices-- demarcate the dry valleys from adjacent interfluve slopes. Gradients are gentle or moderate and at their mouths these surfaces are coterminous with adjacent campiña valley floors. In some cases, the dry valleys are fed by other dry, flat-bottomed, lower order tributaries.

Prominent examples of flat-bottomed, dry valleys occur at Km 95 and Km 103 of the Madrid-Zaragoza highway incised 50 to 100 m into the Pontian Limestone.[5] These valleys meander broadly before reaching the páramo edge where gradients increase dramatically to join with the campiña floor more than 100 m below. These particular features are relict--perhaps ancient karst depressions--as shown by the regrading of the lower valley reaches to the modern Rio Badiel valley floor.

Arroyo or barranco incision into some flat-bottomed, dry valleys shows that they comprise channel fill and colluvium, rather than bedrock. Presumably, therefore, these are relict arroyo beds that were active during the alluviation cycle of the campiñas to which they are graded. The aperiodically active channels incised into them are graded to the modern river courses. In other situations, however, lateral tributary valleys have built gently concave (8°-20°) alluvial fans onto the campiña surface. Some of these fans also are dissected by aperiodically active drainage channels so that several periods of arroyo cutting may be distinguished.

Karst Features

Although not prominent or widespread, karst features are locally developed in the Hesperian limestone surfaces.[6] In the vicinity of Algora, for example, solution along the strike of folded Mesozoic limestone has effected local relief greater than where the plateau

---

[5]Thornes (1967) described similar features developed in the Pontian limestone near Soria.

[6]Jordana (1935, 50) reported the development of karst topography in the Hesperian Meseta. He described, but did not map, lapiés in the Pontian limestone of Guadalajara Province: "Estos agujeros afectan forman caprichosas y variadas, y hay sitios donde se multiplican tanto, que la roca toma el aspecto de una esponja." In his study of the Soria area, Thornes (1967, 167) comments on the "generally poor development of karst features" while Solé (1952) omits any reference to karst phenomena in the regional treatment of the geomorphology of this area. Aside from these few references, karst features have not been described from the Hesperian Meseta, although Butzer (unpublished) has mapped them in the Ambrona-Torralba area.

surface is comprised of flat or gently dipping limestone (see discussion of comparable developments in Palmer, 1965). This fluted surface bears similarity to the campos de lunar or torcal terrain described by Spanish authors (see de Novo and Chicarro, 1957:63). More common are the solution depressions developed in the subhorizontal Mesozoic bedrock forming the tabular plateau surface. These dolines (dolinas) are shallow sinks usually less than 5 m deep and 25 m in diameter. Occasionally, such as on the páramo north of Horna, the dolines are large enough to be cultivated. A smaller variety of doline development is recorded near Torralba del Moral, above the Torrente de Sahuco at 1146 m. Solution of carniolas limestone has formed an inverted cone more than 13 m deep and 60 m in diameter at its rim.

Karst features of the Pontian limestone surfaces are limited to small (about 10 m in diameter) shallow solution pans and local lapiés surfaces. The sinks are subtle relief forms not always apparent in the field but readily identifiable on aerial photographs as darkened, symmetrical, drainage depressions. Elsewhere, such as northeast of Torija, the Pontian limestone surface is characterized by large, shallow (20 to 40 m) bowl-like depressions up to 2 km in diameter. These features lack exterior drainage and derive from large scale solution. Lapiés surfaces may be found in Pontian páramo limestone surfaces that are locally bare of terra rossa soil. The exposed limestone is pitted and etched by solution but the surface features rarely exceed 30 cm in relief (see Fig. 6).

The contemporary semi-aridity of the Hesperian Meseta and restrained present rates of solution suggest that these are relict features. The underdeveloped or incipient form of the limited karst phenomena indicates further that the period of karst development probably was comparatively brief. Mensching (1956) recorded relict caves and polja developed in the Jurassic limestones on Mallorca in modern climatic and topographic contexts comparable to those of the Hesperian Meseta. He held that the features date from pre-Würm humid periods and perhaps even from the Tertiary. The cooler and more moist phases of the Pleistocene are the more logical periods of karst development in the Hesperian Meseta. Such formation must have succeeded the Pliocene planation of the Pontian limestone surfaces on which they occur (see following chapter), and the entrenchment of the major Alto Henares valleys with their Plio-Pleistocene features.

Résumé

Distinctive types of land-surface form are identifiable within the meseta morphologic province. In addition to the tabular form of the páramo or plateau surface, paramera, serrania and campiña terrain classes are recognized. Two other features of the land-surface form, flat-bottomed, dry valleys and karst phenomena, can be noted.

The surface configuration discussed in this chapter is etched into the geologic stage reviewed in Chapter II. The bio-climatic constraints examined in Chapter III define the contemporary geomorphic setting. Two elements of the land form are of particular interest because they describe major components of the land-surface form: the páramo surfaces, and the Pleistocene features of the campiña valley bottoms. These topics are the subjects of the subsequent chapters.

PART II

THE UPLAND SURFACES AND LOWLAND PLEISTOCENE FEATURES

CHAPTER V

UPLAND EROSIONAL SURFACES OF

THE HESPERIAN MESETA

Introduction

The term _mesa_ is used widely to designate flat-topped hills preserved by subhorizontal sedimentary strata. Reasonably enough, the term derives from the Spanish
word for "table"; less logical, however, are the structural connotations frequently implied
by its usage. In the Spanish literature, _mesa,_ or its augmentative _meseta_, is used in reference to extensive high areas commonly planed by erosion (see de Novo and Chicarro,
1957:22, 214; also Solé, 1952:205) much in the same way that "plateau" is used in the
English literature. [1]

The effect of planation on upland surface form is easily overlooked. Proceeding
northeast from Madrid on Carretera Nacional II, for example, high mesas and castellated
escarpments fringe the lower Henares valley between Torrejón de Ardoz and Guadalajara.
The colorful variety of the subhorizontal Tertiary deposits comprising these uplands dis-
guises local evidence of planation, so much so that structure might be considered the factor responsible for development of the tableland surface form. But closer observation
shows that these uplands frequently truncate inclined strata to indicate that the tableland,
and by implication the entire Meseta, is a planation surface conforming to the implicit
Spanish meaning of the term _meseta_. Structural control or influences are not to be
ignored but such observations demonstrate that the uplands of this part of the Henares
basin as well as of the Hesperian Meseta are, in fact, broad planation surfaces.

The sweeping vista of undulating plain viewed from the edge of the upland rim at
Torija offers additional deception. The accordance of horizons provides the illusion of a
single extensive surface, broken only by the distant Cordillera. However, a rather different impression results from within the valleys that are incised some 200 m into this surface by the headwaters to the Rio Tajo. The degree and intensity of dissection become

---

[1]Of note is the fact that of the major introductory geography texts in the U.S.A.,
only one (Kendall, Glendenning and MacFadden, Introduction to Geography, Harcourt,
Brace and World, 1962:183) relates mesa or meseta land-surface form development to
planation processes.

apparent here and several distinct planation surfaces can be recognized. Identification and development of these surfaces as they occur within the Alto Henares are the focus of this chapter.

Previous Research

Earlier studies refer to erosional surfaces in the Tajo Basin and the Hesperian Meseta (see Lotze, 1929; Hernández-Pacheco, F., 1932; Richter and Teichmüller, 1933; Birot, 1933), but Julius Schwenzner (1937) was the first to undertake a systematic study of the upland surfaces. Schwenzner spent more than five years studying the geomorphology of the eastern sector of the Cordillera Central and the adjoining Mesetas, including the Hesperian Meseta, and identifying and mapping the multiple planation levels. The monograph that resulted remains unsurpassed in scope of inquiry and thoroughness of detail and his observations generally hold up well under field examination. The following synopsis, as well as the frequent references throughout this study, reflect the importance of his work.

Erosional Surfaces According to Schwenzner[2]

Dachfläche

The oldest planation surface recognized by Schwenzner is the mid-Tertiary peneplane (intratertiäre Rumpffläche) recorded in the modern Meseta landscape by isolated remnants (Dachfläche, "summit surface"). Such inselbergs are offset 50 to 100 m above the general elevation of the páramos. Pico de Ministra (1309 m) and Loma San Sebastián (1289 m) along the Tajo-Ebro watershed, and Loma San Cristobal (1213 m) southwest of Algora are three such remnants in the study area (pp. 25, 26, 66). Schwenzner notes that the mid-Tertiary (Savian) orogeny followed the cutting of the Dachfläche but was completed by the time of the Sarmatian-Pontian transition so that at best he is able to place its age between the early Oligocene and the Tortonian (pp. 112, 117). The Oligo-Miocene disconformity revealed in some of the Tertiary basins is also related to this erosional period (pp. 42f).

In the Somosierra, the D-surface is confined to uparched ridges and crestlines of the sierra between 1550 and 2200 m. Where the subsurface comprises brittle crystallines, block faulting occurred in contrast to the corrugated undulations and folds of other areas. Elevations of the Dachfläche in the west of the Somosierra lie between 1700 and 2150 m

---

[2] All references in this section are to J. Schwenzner, "Zur Morphologie des Zentralspanischen Hochlandes," v. 10, 1937. References in the text are to page numbers in this monograph.

and at 1800 to 2100 m in the east (p. 93). Overall elevations decrease from 2000 to 2150 m in the northern crest line to 1700 to 1800 m in the south (p. 99). The Sierra del Alto Rey, which mark the eastern margins of the Somosierra immediately northwest of the study area, records the D-surface at 1760 to 1834 m (p. 94).

Meseta Surface 3

The modern landscape (Landschaftsbild) of the Hesperian Meseta is dominated by the "Meseta Surface 3" (M3) occurring at 1150 to 1250 m. This feature is best developed in the eastern sector, though it is also conspicuous within the study area where little deformation of the Hesperian block has occurred. Younger tectonics have dislocated segments of the M3 so that presently it is found at 1300 to 1350 m along the eastern Hesperian saddle, and between 1400 and 1500 m along the unwarped and upfaulted margins of the Somosierra (pp. 65, 67, 71, 96ff). Within the Somosierra proper, the M3 is preserved by surfaces at average elevations of 1550 to 1600 m and 1440 to 1450 m in the northern and southern foothill regions respectively (p. 99). According to Schwenzner, continued uparching of the M3 during the late Tertiary was responsible for the fact that the margins of the M3 surface may dip beneath Mio-Pliocene beds at 1150 to 1170 m (pp. 66ff).

Conglomeratic deposits derived from Pontian páramo limestone on top of the M3 and dislocated by post-Pontian tectonics are believed to demonstrate that most of the planation of the M3 must have preceded the Pontian (pp. 62f, 65, 68).[3] However, Schwenzner unfortunately provided grounds for confusion by referring to the M3 surface not only as the Pontian peneplane (see pp. 63, 72, 82, 89, 95, 102) but also as the erosional correlative of the Pontian limestones (see pp. 114, 117). There should be little doubt, however, that throughout his study the M3 surface is essentially but implicitly considered as a pre-Pontian feature.

Meseta Surface 2

The "Meseta Surface 2" (M2) is younger than the M3. It is extensively developed only among the northern and southern piedmonts of the Cordillera Central where it is preserved along the margins of the Castilian basins between 1000 to 1150 m, generally on late Tertiary deposits. These are basically considered as pediment surfaces (Felsfussflächen, p. 118), frequently covered by veneers of quartz and quartzite conglomerate, locally increasing to 30 m in thickness near Hiendelaencina (p. 106). Small inselbergs

---

[3] " . . . so muss die Bildung der M3-Fläche bereits zu Beginn des Ponts im wesentlichen abgeschlossen gewesen sein" (p. 63). And again, " . . . die M3-Fläche, deren Bildung im Pont beendet ist" (p. 71).

occasionally rise 20 to 50 m above the general 1050 to 1100 m level of the M2 in the Hesperian Meseta (pp. 61, 72). In other areas, the M2 is preserved as valley bottoms cut
into the higher M3 surface (pp. 69f). Perhaps the most conspicuous expression of this erosional level is the páramo of La Alcarria (750-1100 m). A post-Pontian age is assigned
to this bevelling since the M2 cuts across deformations in the Pontian limestones (see
pp. 43ff).

Meseta Surface 1

The youngest erosional cycle is represented by the M1 surface, prominent as terraces in the Tertiary basins and as valley bottoms in the Hesperian Meseta. Bevelling
was thought to have begun in the centers of the great tectonic basins, while the surfaces
were gradually extended up-valley in the form of fluvial platforms (Flurterrassen). The
M1 in the center of the Tajo Basin is now found at 650 to 700 m elevation and covered with
3 to 5 m of well-stratified, quartzite and limestone conglomerate with a yellowish-red
matrix (Schwnezner, 1937:47). North of Madrid, the M1 dominates the New Castilian piedmont at 880 to 950 m under fanglomerates known as rañas. In the Hesperian Meseta,
Schwenzner (p. 70) considered the headwater basins of the Rio Salado (1020-1050 m) and
Rio Cañamares (1080-1110 m) as well as terraces at Riba de Santiuste (1000-1020 m) and
Atancé (990-1010 m) as expressions of the M1 surface. He observed that flexing of the
Hesperian Mass effected the M1 much as it had the M2 and M3. Development of the M1
postdates the M2 and consequently was thought to be of Plio-Pleistocene age.

Phases of deformation

According to Schwenzner, each of these three cycles of erosion was preceded by
deformation and/or orogeny which provided the potential energy necessary to initiate each
subsequent period of planation. So, for example, the major mid-Tertiary orogeny, the
Savian phase, preceded planation of the M3 (pp. 71, 97). Updoming and warping of the
Sierra de Guadarrama and Somosierra, by as much as 600 m, and downwarping of the
Tajo and Duero Basins established a differential relief in the order of 1200 m (p. 117).
Faulting within the Hesperian block may also be related to this general period (pp. 72, 89).

A second Tertiary period of folding, faulting and tilting intervened between the planations of the M3 and M2 surfaces. The Somosierra experienced faulting in the order of
200 to 250 m (pp. 67f, 97f, 99) and dislocation of the M3 surface along the northern edge
of the Sierra de Guadarrama attained 350 m (p. 101). In addition, Schwenzner recognized
uparching of the Hesperian saddle in the order of 200 m (pp. 68, 72) and concluded that
the two general levels of the M3 in this area can be attributed to such flexing (see pp. 64-
68). This period of orogeny, the Rhodanian, is held to be of post-Pontian age, consisting

of two subphases. In contradistinction to Schröder (1930, 178), Schwenzner (1937, 71) held that this was a period of major faulting and flexing in central Spain.

Following the M2 planation, gentle late Pliocene deformation upraised the mountain blocks by some 100 to 120 m with some attendant warping of Tertiary basin deposits (pp. 115, 118). And finally, deformation of the M1 surface, development of which has presumably continued to the present day, extended through the Pleistocene as is indicated by dislocations of the high fluvial terraces recognized by Schwenzner (pp. 118f) as basin counterparts to the headwater valley expression of the M1 surface.

In overview, Schwenzner's scheme of morphological evolution encompasses three cycles of erosion beginning during mid-Tertiary times. This sequence is illustrated in Figure 13. Planation of the Dachfläche, dating between mid-Oligocene and early Miocene times, was interrupted by the mid-Tertiary (Savian) orogeny which, by implication, marked the early Miocene. This paroxism provided the potential energy for planation of the M3 surface, completed by Pontian times. After this, renewed deformation (Rhodanian) initiated a new phase of planation leading to development of the Pliocene M2 surface. Finally, late Pliocene to early Pleistocene uplift led to valley downcutting followed by restricted planation of the M1 surface, locally superseded by younger riverine terraces that developed in Pleistocene times as a consequence of tectonic adjustments.

Erosional Surfaces According to Solé and Birot

L. Solé Sabarís (1952, 171f; 188-90; 215-17) dismisses the scheme proposed by Schwenzner in favor of one he considers simpler and more complete. Integrating evidence from several geomorphic provinces--the Meseta of Extremadura, the Castilian Mesetas, the Cordillera Central and Cordillera Ibérica--he concludes that there were but two erosional cycles of Tertiary age. The following evolution is envisaged (see Fig. 14):

1) Pre-Miocene (presumably Oligocene) Savian orogeny uplifted and deformed the Paleozoic basement, creating the Central Sierras; lateral pressures folded the Cordillera Ibérica and produced the north-south folds of the eastern Somosierra. The Castilian Basins were initiated at this time.

2) Early Miocene ("pre-Tortonian," p. 188) Styrian orogeny produced the broad arching of the Cordillera Central, with major faults that delimit the exterior margins of the Cordillera block. Erosion removed much of the early Tertiary mantle within the Cordillera and partially exhumed the ancient (pre-Cretaceous) erosional surface while filling in the graben troughs of the mountain block.

3) The mid-Miocene sedimentation cycle is recorded by the Tortonian and Sarmatian conglomerates, sands, marginal clays, and lacustrine limestones, gypsum and marls. Sedimentation burying the earlier structural topography continued throughout the Miocene as a result of the erosion that continued to the end of the Pontian and that cut the "fundamental" meseta peneplane (Penillanura fundamental). This is referred to as the first, or upper-level, erosional cycle.

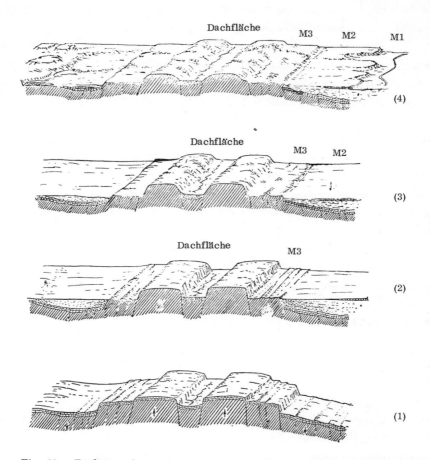

Fig. 13.--Evolution of erosional surfaces according to Schwenzner. 1) Interruption of mid-Oligocene to early Miocene planation (<u>Dachfläche</u>) by the Savian orogeny.
2) Planation of the M3 surface (mid-Miocene) with <u>Dachfläche</u> residual hills. 3) Rhodanian orogeny followed by planation of the M2 surface during the Pliocene. 4) Late Pliocene to early Pleistocene M1 planation and development of river terraces.

4) Post-Pontian (early Pliocene) uplift rejuvenated the sierra, deforming the meseta peneplane. This "third orogenic crisis" (p. 188), the Rhodanian, is attributed to isostatic adjustment.

5) Late Pliocene arid-tropical conditions ushered in a second planation cycle, associated with deposition of the <u>rañas</u>. This surface may cut across inclined limestones of the Pontian, across faults that make the Tertiary/Paleozoic contact in the piedmont zone, and across undulating Miocene and inclined Cretaceous beds. Preserved at elevations of 1000 to 1100 m in the southern Meseta, this erosional surface locally is capped by 3 to 5 m of yellowish-red quartz and quartzite gravel (p. 216). This is referred to as the second or lower-level erosional cycle.

63

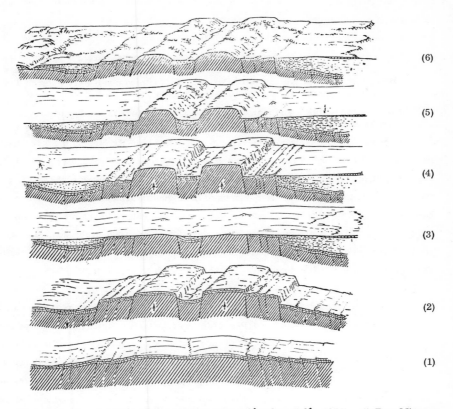

Fig. 14.--Erosional surfaces according to Solé (after Solé, 1952). 1) Pre-Miocene dislocation with incipient formation of Cordillera Central and Castilian basins. 2) Early Miocene (Styrian) orogeny with major faults delimiting the Cordillera Central. 3) The first erosional cycle (mid-Miocene); cutting of the Penillanura fundamental. Castilian basin sedimentation of the Miocene lacustrine and riverine facies ending with the Pontian limestone. 4) Post-Pontian (Rhodanian) rejuvenation with folding of edges of Tertiary Basins. 5) The second erosional cycle (Pliocene) forming Cordillera Central pediment ramps and inselbergs. 6) Post-Pliocene warping with development of modern hydrography and dissection of the páramos.

6) The Quaternary was characterized by further warping as well as the development of tectonic and climatic riverine terraces (p. 217).

In a later, more exhaustive and remarkably conjectural study of the Cordillera Central, Birot and Solé (1954) further developed the viewpoint that only two erosional cycles were responsible for the multi-surfaces of the Meseta. The authors are struck by the juxtaposition of crystalline masses crowned by surfaces and delimited by rectilinear scarps, and of extensive surfaces of erosion with inselbergs that are undeformed. After lengthy considerations, a scheme of evolution is only suggested, since there is an "absence of available data as to the intensity of faulting" (p. 77):

1) Deformation throughout the early Tertiary.

2) Major mid-Miocene movement, culminating a general period of Miocene instability (p. 48) that uplifted the central Guadarrama horst and concluded in Pontian times. Simultaneously the Cordillera scarps and piedmonts were subjected to erosion, creating pediments with inselbergs (Cycle one).

3) Pliocene stability and planation, punctuated by uplift in the Cordillera, that have continued into the Quaternary (Cycle two).

According to this scheme the various surfaces within the Meseta must be attributed to continuing Mio-Pliocene deformation and to subsequent isostatic adjustment. So, for example, Birot and Solé (1954) relate the Pontian páramo (La Alcarria), the intra-Cordilleran Pontian surfaces, the surfaces of the Sigüenza area and those near Campisá-balos (1400 m) to the first or Miocene cycle. The last locational example is thought to be tectonically offset from the generally lower elevation of this first cycle surface. The second Pliocene erosional cycle would have produced surfaces at Hiendelaencina (1080-1140 m), Riaza, and on the interfluves of the Duero left bank tributaries. The rañas relate to this Pliocene cycle.

In the way of explanation, Solé maintains much as he did in his previous study that the two erosional levels relate to periods of tectonic instability during the Miocene and Pliocene. Birot, on the other hand, is of the opinion he too had formulated in an earlier study (Birot, 1933) wherein deformation in the Cordillera Central and Meseta derives from the Miocene only. This phase resulted in the development of the older, higher erosional surface. Since, as Birot maintains, there has been no major tectonic movement since the Pontian planation, the lower Pliocene erosional levels and Cordillera pediments are attributed to climatic change at the beginning of the Pliocene.

Aside from these differing explanations and the few specific, locational references for the two surfaces, Birot and Solé (1954) offer no additional insights with regard to the study area. It is regrettable that the two very generalized maps of the Meseta surfaces included in this study are of little relevance in attempting to assess their conclusions.

Erosional Surfaces of the Hesperian Meseta

Mapping Procedure

Schwenzner's large scale map (1:400,000) of the Meseta surfaces introduces generalization that eliminates local detail which in particular instances may have relevance for the evolution of planation surfaces. Because of the different hypotheses of genesis, a more careful investigation of the upland surfaces within the study area seemed warranted and was carried out. The results proved to be useful and meaningful for several reasons.

In the first place, faults and deformation of this sector of the Hesperian Meseta had already been well mapped and studied.[4] Consequently, tectonic dislocations could be eliminated as explanations for the origin of the multi-level surfaces of the Hesperian Meseta, at least within the area studied. Second, since the study area encompasses the headwaters of three major drainage basins--Duero, Tajo and Ebro--incision rather than planation has generally been dominant. Consequently, the association of paleo-surfaces is better preserved than in the middle and lower sectors of the basins of these rivers, where large-scale meandering has tended to undercut and destroy vestiges of older planation surfaces. And, finally, the lithologic variety of the Hesperian Meseta provides a unique opportunity to examine the relationship of the several surfaces as they have developed and cut across the entire range of the Mesozoic and Tertiary geologic foundations. These factors justify the detailed re-mapping of the erosional surfaces undertaken for the study area. They also emphasize the relevance of any conclusions reached from this effort. In view of their relevance, the methods and procedures used can be outlined here.

Initial identification of a particular surface was made in the field. Higher elevations served as vantage points from which broad surfaces or platforms could be distinguished. The "flatness" of the meseta panorama apparent from these vistas accentuates any vertical deviations of land-surface form. So, for example, interruptions of accordant horizons draw the eye to higher remnants, while lower surfaces are readily identified from the perspective of higher observation points (see Fig. 15). Observations were then recorded on the 1:50,000 topographic maps (20 m contour interval) and final demarcation of the platform surfaces was made from stereographic pairs of aerial photographs. This procedure assured that identification of a given surface derived from first-hand field observation, unprejudiced by pre-existing maps or research.

By far the more difficult and critical phase of the actual mapping was to interrelate the many discontinuous and non-adjacent mapped surfaces, in order to reconstruct the former extension of contemporary planation surfaces. The pieces of this puzzle were the disjunct surface segments that comprise the general accordance of a surface horizon viewed in the field panorama. Two factors complicate this procedure: 1) The planation surfaces in the Hesperian Meseta lack characteristic deposits so that the usual lines of stratigraphic reasoning are not available. Schwenzner (1937, 62; 106) had, indeed, identified post-planation deposits on some of the Meseta surfaces but, aside from the fact that none can be recognized within the study area, these are not significantly diagnostic for

---

[4]See, for example, Schwenzner (1937) Map 2, and the relevant memoirs of the Mapa Geológico de España, including Sigüenza (No. 66), Barahona (No. 67), Ledanca (No. 158) and Maranchón (No. 285). More recent 1:25,000 preparations of the Sigüenza sheet (461) by the Instituto Geología Económica (in press) are also useful.

Fig. 15.--Erosional surfaces west of Sigüenza. The dissected anticline near Sigüenza occupies the foreground beyond which lies the town. The B, C, and D surfaces are represented by the general level of the páramo and the residual hills above and below this level. The Cordillera Central are in the distance.

stratigraphic correlation. 2) Disjunct surface remnants cannot be correlated solely on the basis of absolute elevation. Evidence for this is provided by the present day campiñas, some of which Schwenzner (1937, 70) considered as incipient erosional surfaces. Local relief in these situations may attain 40 m or occasionally perhaps even more, but any attempt to delimit different erosional surfaces on this basis would be erroneous.

In consequence of these general qualifications the best procedure to correlate discontinuous segments of paleo-surfaces hinges upon those instances where vertically offset planation platforms occur immediately adjacent to one another.[5] Since, as previously indicated, tectonic adjustments can be discounted within the immediate study area, such occurrences best demonstrate the existence of multi-levels. Other surface segments mapped in the field were related to such suitable locales either directly or indirectly by using intervening expressions of the same surface. Most of the surface segments identified in Figure 16 were traversed at least once during the field season; from each vantage point the broad scale inclination of each planation surface was noted and taken into account when relating isolated surface segments. In this way most segments were studied from more than one perspective, so that distortions of depth perception could be avoided. Distant associations were appraised in the field by Abney level, corroborated by map and aerial photograph study.

Treatment of Surfaces

Except for the extensive Pontian limestone páramos, planation surfaces were not mapped in areas of Tertiary rocks. Platforms occur locally in the land-surface form, such as that west of Santiuste (1030-1060 m) or the interfluve between the Rio Henares and Rio Dulce (940-980 m), but as often as not these relate to the resistant, massive conglomeratic strata within the Oligo-Miocene sequence. Remnants of other planation surfaces undoubtedly exist within the area of Tertiary basin sediments but their infrequent occurrence and associated structural factors would make the mapping of surfaces highly circumspect. Platform surfaces within the Mesozoic uplands have not been mapped if structural factors seem to assume dominance. So, for example, flattened knolls within the cores of the truncated anticlines near Riba de Santiuste and Sigüenza have not been mapped as surfaces even though there may be a striking accordance of summits.

Four distinct surfaces can be recognized within the study area on the basis of field reconnaissance and mapping. These are hereafter identified as the A, B, C, and D surfaces, from oldest to youngest, and are mapped in Figure 16. The ultimate treatment of some surface segments proved sufficiently problematic to warrant comment or discussion. The following subsections relate to these areas and implicitly demonstrate aspects of the overall mapping procedure.

Upper Dulce drainage

An informative complex of planation surfaces is recorded along the upper Dulce headwaters in a suite traversed by the Carretera Nacional II (see Fig. 17, profile 3). At

---

[5] Prominent examples of these situations occur immediately east of Sauca, along the flanks of the upland of the upper Rio Salado, and on the western flank of Sierra de Ministra. Refer to the profiles in Figure 17.

68

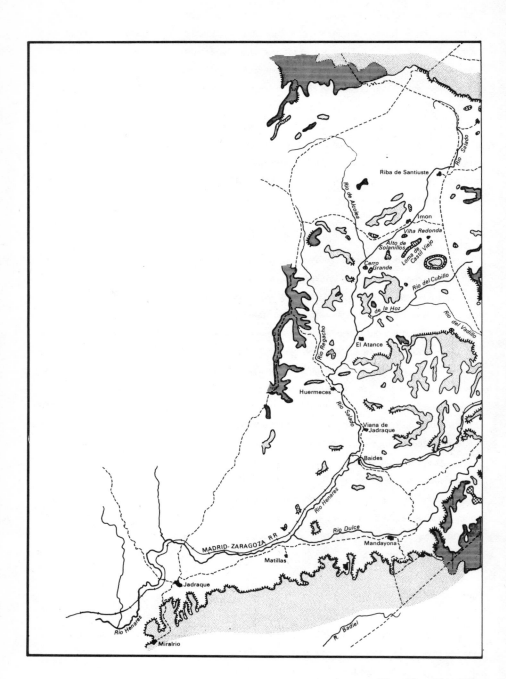

Fig. 16.--Planation

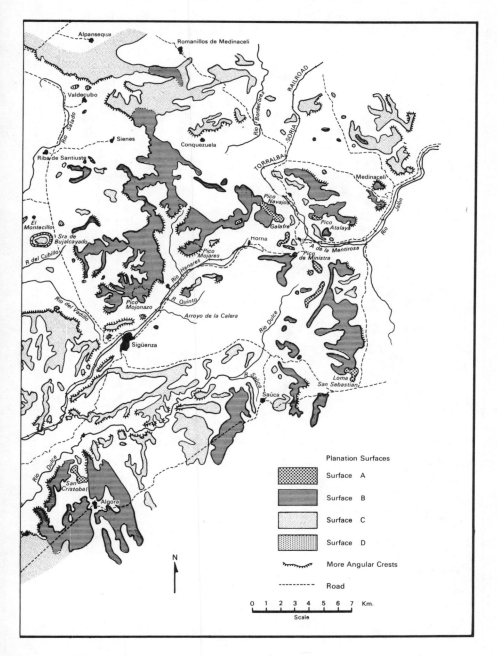

surfaces of the study area

70

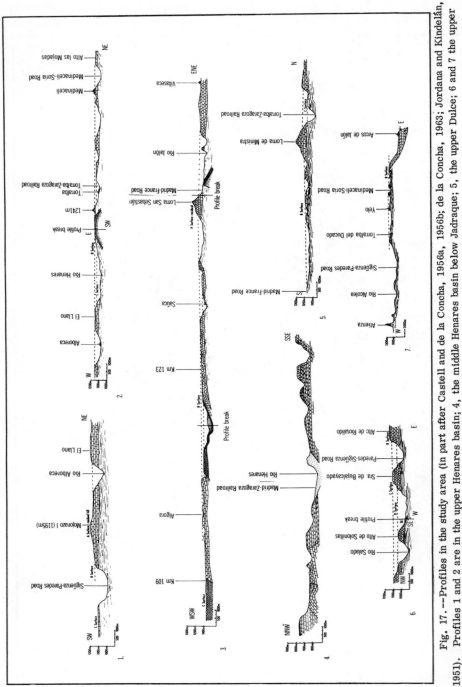

Fig. 17. --Profiles in the study area (in part after Castell and de la Concha, 1956a, 1956b; de la Concha, 1963; Jordana and Kindelán, 1951). Profiles 1 and 2 are in the upper Henares basin; 4, the middle Henares basin below Jadraque; 5, the upper Dulce; 6 and 7 the upper Salado; and profile 3 traverses the Henares-Jalón drainage divide.

Km 109 (1080 m) the roadway begins to surmount an upland ridge that reaches from Loma San Cristobal (1213 m) southeast to the Sierra Megarrón (1250 m). This structurally preserved feature, within which the village of Algora nestles, is the "B" surface. Shortly beyond Km 115 (1140 m) it gives way to a campiña developed in weaker, more erodible red sands above which Liassic limestones have preserved remnants of the "C" surface. Between Km 121 and 123 the "C" surface, dominant throughout the upper Dulce, is traversed by the highway at about 1140-1160 m before once again ascending onto the "B" surface at Km 125.4 to 126 (1170-1180 m). Descent into the Rio Saúca valley at this point shows a lower "C" platform along the valley margins, a remnant of which the highway traverses between Km 129.3 and 129.7 (1150-1160 m). By Km 130.8 a higher "B" surface ridge (1200 m) is crossed as the highway winds briefly among upland expressions (Km 131.4, Km 133) of this same surface (1190-1210 m).

The transitions from the Pontian limestone páramo "C" surface to the higher "B" surface at Km 109 and again at Km 123 are gradual but nonetheless apparent and confirmed by the aerial photography. The prominence of structural controls in this area cannot be overlooked where lineation of the higher "B" surface ridges reflects the upturned edges of Cretaceous limestone.

Schwenzner had correctly recognized Loma San Cristobal as a Dachfläche remnant. Its shoulders are here considered as remnants of a planation cycle intervening between the cutting of the "A" and "C" surfaces, although an alternative interpretation might hold these levels to be degraded remnants of the higher "A" surface. This would require, however, that the long phase of "B" planation, well recorded elsewhere in the Alto Henares, be ignored at this point. Consequently, these lower flanks of Loma San Cristobal are more reasonably considered representative of the "B" surface, although modified by structural features.

Upper Henares drainage

The Sierra de Ministra is the keystone of the interfingering of the surfaces of the Alto Henares since evidence of three planation surfaces is preserved here. The highest peaks (1309 m and 1289 m) are remnants of an "A" surface; the saddle between these at 1200 to 1235 m and the several interfluve spurs comprise the headwater rim of the "B" surface, deeply dissected by the modern campiñas. East and north of Sierra de Ministra stretch the páramos of Old Castile while those of New Castile unfold to the west and south. In addition to establishing the Duero-Ebro-Tajo watershed, the Sierra de Ministra marks the interdigitation of surfaces of Old and New Castile, thereby allowing correlations of the several meseta surfaces across the drainage divide.

Upper Salado drainage

The smooth rise between the "C" and "B" surfaces north of Paredes de Sigüenza is subtle so that its demarcation is somewhat inferential, although the existence of two levels along this northern rim of the Salado drainage is not in doubt. Field and aerial photo study confirm this. Additional corroboration is attained by tracing the segments of the vertical break in question through their respective surficial counterparts to the more clearly off-set surfaces found adjacent to the Rio Masegar valley. Distinction of the "B" and "C" surfaces in this area is significant since they are developed in the same lithologic unit (Rhetian limestone) and a structural origin is precluded.

A surface lower than the extensive "C" surface is suggested throughout the upper Salado basin by high entrant platforms, preserved along the flanks of the páramo surfaces, and by inselbergs or hills (cerros) that interrupt the modern campiña. This is the "D" surface.

Schwenzner (1937, 70) identified the broad valley bottom of the upper Salado at 1020 to 1050 m as an erosional surface that forms a terrace at 1000 to 1020 m east and south of Riba de Santiuste. Since the actual valley bottom of the upper Rio Salado is 20 to 30 m lower than stated by Schwenzner, question arises as to whether he meant the campiñas of the Salado or the marginal 1020 to 1050 m entrant platforms. In any case the "D" surface as mapped in Figure 16 does not relate to Schwenzner's terraces nor to the widespread terraces of the upper Salado that are below and stratigraphically younger than these "D" surface features (see Chap. VI).

The "A" surface

The oldest erosional surface is recorded in remnant form by inselbergs that break the otherwise flat meseta upland. These features generally exceed 1220 m in elevation and are preserved most prominently along the Ebro-Tajo-Duero drainage divide: Pico Navajos (1236 m), Pico Galafre (1231 m), the summit level due south of Torralba del Moral (1241 m), Pico Ministra (1309 m) and Loma San Sebastián (1289 m). Other remnants are found northwest of Medinaceli (Villanueva, 1245 m), northeast of Torralba del Moral (Atalaya, 1239 m) and north of Algora (Loma San Cristobal, 1213 m). In addition to these features, subdued residuals rising above the younger "B" surface may suggest more degraded relics of this same "A" surface. Examples are recorded southeast of Alboreca (Mojares, 1196 m) and west of Alcuneza (Mojonazo, 1195 m). The "A" surface is equivalent to the Dachfläche identified and mapped by Schwenzner (1937, Map 3).

The "B" surface

Generally occurring between 1100 and 1235 m, segments of the "B" surface are found throughout the Hesperian Meseta. Between Mirabueno and Saúca, it is planed across strongly deformed and inclined (40° near Algora) Cenomanian and Rhetian limestones, the latter dipping as much as 20° west of Alcolea del Pinar. In the former instance, solution along major bedding planes has altered this surface by producing linear drainage lines, such as in the vicinity of Algora. Those surface segments which cut across the Rhetian limestone show incipient, dendritic drainage lines, gently incised some 10 to 20 m into the limestone.

West of the Sierra de Ministra watershed, the "B" surface is bevelled to the southwest; to the east its extent is limited so that an overall eastward inclination can only be suggested. These counterparts of the "B" surface abut one another in the saddle of the Sierra de Ministra where the headwaters of the Rio Jalón and Rio Dulce are consuming the section of the "B" surface that forms the Ebro-Tajo watershed.

The uplands comprising the "B" surface are not as featureless as the Pontian limestone páramo surfaces. Greater local relief derives from extensive solution of the various limestones; low-order drainage lines sink abruptly 30 m or more into the surface to form initial swales of convex-concave shape. Where the swales assume more extensive development, they feed channels so deeply incised into the upland edges that they affect the irregular configuration of the surface segments shown in Figure 16. Crest slopes are commonly subrounded and give way to midslopes of 20° or less. Angular to subangular crest slopes occur where fluvial incision, relating to the present campiñas, has increased the local relief and favored rapid undercutting. Such cases occur, for example, above the Rio Dulce gorge, near Pozancos in the upper Rio de la Hoz, and in the upper Rio Henares. In each of these cases the present river lies 100 to 160 m below the shoulders of the "B" surface. The well-developed paleo-solution features and the rounding of crest slopes measurably reflect the great age of the "B" surface.

The angular-subangular limestone detritus ($<$10 cm) that litters the thin xerorendzina (Chap. III) mantle of the "B" surface cannot be considered as peculiar to this surface. This detrital material is common to all undisturbed limestone terrain within the study area. Such surficial debris can be attributed to frost-weathering and is related to the Pleistocene breccias now encountered over much of the Hesperian Meseta. Schwenzner (1937, 62; see also Schröder, 1930:159f) identified conglomerates on the Barahona-Paredes and Barcones-Romanillos uplands which he claimed derived from the degenerated Rumpffläche. This author was unable to locate such conglomerates in the field, but it is likely that Schwenzner referred to the conglomeratic tufa that is characteristic of certain lower Pleistocene deposits nearby (see Chap. VI). It would, therefore, appear that the "B" surface lacks diagnostic deposits.

The "C" surface

Precise delineation of a boundary between the "B" and "C" surfaces is confounded by subtle transitions. Unmistakable vertical offsets do occur, such as above Saúca (Fig. 18) and in the Rio Masegar valley, so that distinct planation levels cannot be ignored. But elsewhere the distinction is less clear and can be identified only after careful field observation. This abstruseness derives from the basin-head disposition of these surfaces, where destruction of the "B" surface would have been restricted because of limited relief energy and the subsequent degradation which further smoothened the offset.

The expansive Pontian limestone páramos and parameras of the Castilian Mesetas are the most notable expression of the "C" planation surface. The páramo of La Alcarria abuts the Hesperian Meseta near Mirabueno and stretches to the southwest where its ramparts (750 m) overlook Madrid. It is the land-surface form of these tablelands that comprises the monotonous, featureless landscape of the meseta. Except where low-order streams breach the edges of the limestone cap, organized drainage patterns appear to be absent on the surface; most drainage is internal to the large but very shallow depressions previously described (Chap. IV). As a consequence, local relief is negligible and unapparent, frequently being less than 1 m per 100 m square. The 1:50,000 topographic series shows several examples of 1 km squares void of relief where 20 m contour intervals are used. The monotony of these uplands is sustained further by the infrequent occurrence of woodland; the 1 to 2 m stands of matorral are broken only occasionally by more dense, 2 to 5 m tall groups of live oak. Cultivation of the terra rossa mantle is limited to wheat and barley, reflecting the edaphic and climatic aridity.

Northeast of the Sierra de Ministra watershed and immediately adjacent to the study area, the Páramo de La Mata constitutes the "C" surface counterpart of La Alcarria for the Duero and Ebro Basins. This La Mata surface is also cut across Pontian limestone bevelled toward the Rio Duero Basin at 1100 to 1160 m.

Within the Hesperian Meseta proper, the "C" surface interfingers with "B" and "A" surface remnants. In the upper Dulce and middle Henares basins it cuts across Rhetian, Charmutian, and Cenomanian limestones at 1100 to 1150 m. In contrast to the peripheral páramos, these surface segments are fragmented by the headwater incision of the Rio Dulce, Rio Henares, and Rio Salado. Destruction of the "C" surface in these areas reflects more vigorous stream dissection in response to increases of available relief, resulting from breaching of the intensely folded Mesozoic margins of the Hesperian Meseta by these rivers.

The varied lithologic composition of the "C" surface within the Hesperian Meseta introduces variety to the surficial expression. For example, structural influence upon drainage and surface incision is apparent on the broad upland mass west of Sigüenza. In

Fig. 18.--Erosional surfaces west of Saúca. The horizon formed by the "B" surface is broken by the "A" surface residual hill in the distance. Below the "B" surface a
"C" surface platform is marked by the cultivated fields in the center of the photograph.

the upper Dulce basins lithologic control has allowed low-order drainage lines, such as
the Barranco de la Guardera, to develop gentle, saucer-like valleys in the more erodible
sands and clays of Toarcian age. In these cases the resulting local relief of the more
undulating topography (20-50 m) is greater than in the case of the Pontian limestone
páramo, and the different land-use practices are reflected by the denser and more frequent 3 to 10 m stands of mixed oak.

In addition to these contrasts, development of the "C" surface upon the folded
Mesozoic materials amply demonstrates the planation aspect of this erosional surface.
Throughout the Henares gorge this surface can be seen to cut across Jurassic and

Cretaceous limestones dipping as much as 45°, and in the Dulce gorge across folded strata dipping between 20° and 70°. In the upper Dulce basin the "C" surface is planed across bedrock, dipping at 60° to 80°. Similarly, in the Jalón gorge, east of Lodares, the "C" surface is developed across steeply dipping Rhetian limestones at 1160 to 1170 m. Schwenzner (1937, Map 3) has shown that the Pontian limestones are sufficiently disturbed to demonstrate planation; southwest of Miral Rio the páramo "C" surface cuts Pontian limestones, dipping at 5° to 8°.

The overall slope of the "C" surface is toward the Tajo Basin. Local surface segments, however, are planed toward the major, modern drainage lines. So, for example, the surface remnants in the upper Henares basin show that the "C" surface here was graded to the Rio Henares; comparable situations are evident in the upper Dulce and Salado basins. The upland west of Sigüenza slopes toward the Salado-Henares confluence, also reflecting local conditions of planation. However, in the upper Salado basin the picture is less clear. This relates to the fact that the general elevations of the New Castilian Meseta are lower than those of Old Castile. As a result, the extensive "C" surface of the Barahona upland (1100-1170´ m) developed in relation to the Duero drainage, lies above the general level of the "C" surface (1000-1100 m) of the Tajo drainage.

As in the case of the "B" surface, characteristic surficial deposits are absent from the "C" surface. Schwenzner (1937, 106) recognized gravels on the M2 or "C" surface, but no such deposits were located within the Hesperian Meseta.

The "D" surface

Throughout the headwaters of the upper Rio Henares and its tributaries, isolated mesas and step-like platforms set against the midslopes of the upland surfaces attest to an erosional level lower and later than the "C" surface. In the absence of broad páramo-type mesas or platforms to affirm the development of this "D" surface, its existence might be questioned. However, the repeated occurrence of these features is too suggestive to discount their implications. It is clear that most of the "D" surface has been destroyed during the climatic oscillations of the Pleistocene. Nonetheless, vestiges can be identified between 1040 and 1060 m in the following locations: north of Urés (1050 m), Matas (1050 m), girdling the Sierra de Bujalcayado (1040-1050 m), Montellano (1069 m), Alto de Castro (1047 m), El Montecillo (1040 m), Loma de Castil Vieja (1063 m), Viña Redonda (1053 m), Alto de Solanillos (1046 m), Cerro Grande (1045-1051 m), and platforms north and west of Imón flanking the Llano de las Simas (1040-1050 m). Along the northern margins of the upper Salado basin, platform segments occur at 1080 to 1090 m, such as north of Valdelcubo and west of Madrigal. Additional evidence of this planation surface is recorded in the upper Rio Henares, north of Sigüenza (1060-1070 m) and again north of Estación de Matillas (940-955 m).

The infrequent occurrence of the "D" surface segments and their disposition as valley-entrant platforms point to strict fluvial rather than pedimentation or planation origins, even though alluvial deposits seem to be absent. Such an interpretation probably relegates their age to the Plio-Pleistocene, when the climatic changes peculiar to the Quaternary initiated more vigorous dissection. The cutting of this surface may well have been contemporary with the deposition of the rañas recorded in the interior of the Tajo Tertiary Basin. It should be noted, however, that this is no more than a suggestion since the exact origin of these gravel mantles is still in question. [6]

Résumé of Erosional Surfaces and Comments

The late Mesozoic to early Tertiary earth movements responsible for regression of the Cretaceous seas provided the potential energy prerequisite for the upland erosion and degradation counterpart to the Oligocene detrital sediments that rest conformably on Cretaceous beds north of Jadraque. The great depth of these deposits points to a sustained period of upland denudation of red weathering mantles and regolith, possibly punctuated by periods of greater aridity or of denser vegetation cover, as indicated by the interdigitation of non-clastic deposits in the sedimentary sequence.

This period of general degradation was interrupted by the major mid-Tertiary orogeny which uparched the Hesperian Meseta and severely deformed its peripheries, folding monoclinally the marginal sediments of early Tertiary age along the Cabrera-Baides-Viana de Jadraque-Huérmeces axis. This period of instability is also recorded by the early Miocene (Aquitanian-Burdigalian) hiatus in the Henares headwater area. The great Duero, Ebro and Tajo interior basins were essentially formed at this time.

Waning of the mid-Tertiary paroxism allowed widespread planation to resume dominance at the expense of the deformed Mesozoic uplands of the Hesperian Meseta, cutting the "B" surface of the study area. Residuals of the older, deformed uplands that rise above this surface recall the oldest Tertiary "A" surface. The post-orogenic basin deposits, correlative to this period, are Tortonian ("Tortono-Sarmatian"); they are essentially horizontal and undeformed. The "B" surface is confined to the watershed zone in the study area and nowhere cuts across these mid-Miocene deposits.

Planation of the "B" surface was accompanied by isostatic adjustment in response to unloading. Limited and waning earth-movement may also have occurred as echos of

---

[6] For recent discussions of the controversial raña mantles, see F. Hernández-Pacheco (1962) and I. Asensio Amor (1966), both with references. Several papers of interest are collected in Volume II of the Proceedings of the 16th International Geographical Congress in Lisbon, 1949.

the mid-Tertiary orogeny, but judging by the configuration of the Tortonian beds, they must have been of very limited intensity. This movement nonetheless proved of sufficient effect to impound lacustrine-lagoonal waters in which the massive chalky limestones of the late Miocene (Pontian) were precipitated.

Renewed deformation in late Miocene times (Rhodanian phase) increased the available relief so that the Pontian lakes were drained while erosion and planation resumed on the uplands. This relative movement was sufficient to cut the "C" surface 50 m beneath the prevailing level of the older "B" surface. Planation gently bevelled the subhorizontal Pontian limestones, achieving wide areal expression throughout the Castilian Mesetas. Little remains today of clastic or detrital sediments once associated with this early Pliocene planation; dissolved materials were presumably transported to the sea coast by through-rivers, while clastic detritus would have been reworked along the broad valley bottoms of the Tertiary Basins or deposited as coarse gravels that now overlie the Miocene sequence such as those north of Jadraque.

Final earth movements of limited intensity provided new potential energy during the terminal Pliocene to permit dissection and widespread destruction of the Pliocene "C" surface. Within the Hesperian Meseta this hemicycle of denudation is recorded by "D" surface remnants at 30 to 60 m below the general elevation of the "C" planation surface. More expansive platforms are preserved outside of the study area, along the lower course of the Rio Henares and in the middle Tajo Basin where they are considered to be co-terminal with fluvial terraces of the lower Pleistocene. Cutting of the "D" surface corresponds with deposition of the rañas and other fanglomerates identified in different sectors of the Castilian Meseta. Contemporary with "D" surface development, the Rio Henares began to incise its gorge through the folded Mesozoic beds at the edge of the Hesperian Meseta. Post-Pontian incision is demonstrated by the fact that the present Mesozoic bedrock campiñas lie 30 to 50 m below the "C" surface, whereas below this gorge the Henares valley lies 200 m below the páramo "C" surface. Rock-stratigraphic evidence is lacking but a late Pliocene to early Pleistocene (Villafranchian) time range can be best inferred for planation of the "D" surface.

The scheme of development of erosional surfaces arrived at by this study of the Hesperian Meseta is summarized in Table 3. Of note is the fact that this arrangement is the result of first-hand field observations. Identification, mapping, correlation and ultimate enumeration of these surfaces were performed independent of the geologic history and mapping of the area. The validity of this scheme is, therefore, enhanced by its consistency with the tectonic and sedimentation histories of the study area as determined by other investigators (Chap. II). In order for the discussion to be complete, reference must be made to the geomorphic processes that produced the various erosional surfaces.

TABLE 3

EROSIONAL SURFACES OF THE HESPERIAN MESETA

Modern Surface Form Feature	Upland Erosion	Related Basin Sedimentation
D Surface (valley-side platforms and campiña mesaforms)	Dissection of C surface and planation of D level	Plio-Pleistocene fanglomerates (rañas) (?); valley alluvia
～～～～～～～～～ Isostatic uplift; intermittent warping ～～～～～～～～～		
C Surface (páramos)	Destruction of B surface during planation of C surface	Pliocene non-clastics; local gravel mantles (?)
～～～～～～～～～ Rhodanian deformation ～～～～～～		
B Surface (páramos)	Destruction of A surface during planation of B surface	Tortonian-Sarmatian clastics and evaporites; Pontian limestones during terminal phases
～～～～～～～～ Alpine-Savian orogeny ～～～～～～		
A Surface (residual hills)	General subaerial erosion with development of A surface	Oligocene clastics, evaporites and limestones

Peneplanation or Pediplanation

The examples cited in this text as well as those apparent from comparing Schwenzner's maps of geology (Map 2) and planation surfaces (Map 3) are ample to demonstrate truncation of inclined strata by the erosional surfaces of the study area. Although structural controls may assume dominant roles locally, the planation of these surfaces is nonetheless readily established. Are these peneplanes or pediplanes?

If pediment is taken to mean a rock-cut plain, then these surfaces may be considered as pediments. If, however, by pediment a full complement of associated features is implied, the Hesperian Meseta surfaces are less appropriately referred to as pediments for several reasons. First, the limestone páramos are not strictly piedmont surfaces. Second, sharp breaks in slope (knickpoints) where the surfaces terminate at the mountain footslopes northwest of the study area are obscure or locally absent. Third, alluvial veneers do not occur toward the central depressions of what must have been the ancient internal basins, but now are the extensive páramo surfaces. The raña mantles appear to be confined to the immediate piedmont zones. Finally, pediment form, sensu lato, is universally linked to modern arid or semiarid morphogenetic regions and although the contemporary setting is consistent with this notion, there is no assurance that these conditions prevailed when the relict surfaces were formed.

Are, then, the Hesperian erosional surfaces better considered as peneplanes? This interpretation is less satisfactory since conditions moist enough to permit the fulfill- ment of a "humid cycle" of erosion would also have been sufficiently wet to permit wide- spread solution of the páramo limestone caps. In addition, the commonly cited argument against peneplanation, that the protracted time spans of base-level stability required for nearly complete planation/denudation are in actuality untenable, is no less appropriate here in view of the multi-level, en échelon aspect of the surfaces preserved in the study area. Furthermore, younger phases of denudation/planation that would have produced the lower planation levels at the expense of older, higher ones, must have been accompanied by erosion of the older and higher surfaces. Yet differences in the degree of dissection of the most extensive páramo levels--the "B" and "C" surfaces--are virtually undistin- guishable. Finally, it is widely argued that because of subsequent modification, it is unlikely that Tertiary peneplanes, except perhaps buried ones, have survived to the pres- ent day.

With the above considerations in mind, the question of what geomorphic process produced these surfaces can be addressed. The objections to pedimentation noted above are essentially semantic, while those raised in regard to peneplanation are theoretical. Is there empirical evidence in support of either process?

The Tertiary basin sedimentation that proceeded in conjunction with upland plana- tion (see Table 3 above) provides a record of depositional environments. For the most part the Oligocene and Miocene deposits comprise facies indicative of drier conditions, particularly in the case of the evaporites and limestones. Detrital facies are held to be marginal or peripheral deposits within the larger Tertiary basins (see Chap. II) and may represent the alluvial veneers associated with pediment-cutting. Planation during arid or semiarid periods is favored on an additional, theoretical ground. Regardless of whether lateral planation, backwearing, sheetwash or sheetflood erosion, or a combination of these processes produced the Hesperian Meseta surfaces, denudation of the limestone páramos should have been more effective under relatively dry conditions in which the infrequent and intense periods of precipitation result in abbreviated but effective overland flow. Further, it is supposed that under more humid conditions, sustained overland flow would be absorbed by the limestone bedrock and lead to higher rates of solution and a pro- lific development of karstic features far in excess of the level retained in the contempo- rary landscape.

Finally, previous investigators who have recognized erosional surfaces in central Spain have not considered the processes involved, although most have implicitly assumed some form of pediplanation, since the surfaces are referred to as pediments.[7] However,

---

[7] In his early discussions of the eastern Cordillera Central, Birot (1933, 1945) refers to piedmont pediments developed during arid phases of the late Tertiary. This

the more persuasive argument for the arid or semiarid nature of planation of the páramo surfaces is the very nature of the correlative basin sediments.

Concluding Remarks

In general, the scheme of planation cycles proposed in this study, derived from independent field mapping and checking, agrees with the fundamental findings of Schwenzner. In both cases, four distinct surface levels can be identified on the Meseta: "A, " "B, " "C" and "D" on the one hand, and Dachfläche, M3, M2, and M1 surfaces on the other. General agreement is also reached with respect to the geologic time periods involved; both systems envisage Miocene, Pliocene and Plio-Pleistocene planation of the "B, " "C, " and "D" surfaces respectively, although individual details may differ.

There are, however, divergent interpretations which in some cases reflect significant gaps in Schwenzner's monograph:

1) Schwenzner did not take full advantage of sedimentological evidence in developing his scheme of planation cycles and surfaces. Although relationships of Tertiary planations to Tertiary basin deposits are implied, nowhere does Schwenzner directly correlate basin sedimentation with planation periods. This study relates Oligocene deposits to an erosional period, prior to the major mid-Tertiary orogeny and correlative with the development of the "A" surface. Similarly, planation of the "B" surface is here related to the mid-Miocene (Tortonian) sedimentation, and the "C" and "D" surfaces to the rañas and later phases of fluvial deposition.

2) Schwenzner does not accept a climatic interpretation of fluvial terraces. He relates, instead, the M1 ("D") surface of the upper Henares drainage and the fluvial terrace counterparts in the lower Henares to tectonic impulses during Plio-Pleistocene times. Results of this study show that the "D" surface can logically be relegated to this period but, consistent with the well-developed terrace features of the upper Rio Henares which record Pleistocene climatic changes (see Chapters VI and VII), development of the "D" surface in this area is best related to climatic rather than to tectonic factors.

3) The identification of upper Henares planation surfaces, as mapped in this study, agrees generally with Schwenzner's results. A notable exception, however, is his treatment of the upper Salado "D" or M1 surface. Schwenzner did not recognize any Pleistocene terraces in the campiña of the upper Salado north of Riba de Santiuste and consequently he correlated this campiña with terraces of the lower Henares. A further cause for exception to his interpretation is that the Keuper

---

view is reiterated in his subsequent investigations with Solé (Birot and Solé, 1951, 1954), although these studies and that of Solé (1952) also distinguish the Penillanura fundamental ("peneplane"), which is correlative with the "B" surface of this study. More recently, Sos (1957) considered the raña-veneered surfaces as Pliocene pediments and Mensching (1967) discussed pediments of the northern Cordillera Central piedmont. Schwenzner (1937, 118) also referred to segments of the southern piedmont as pediments. The pediment surfaces referred to in each of these studies occur in the immediate piedmont margins of the Cordillera Central, cut across exhumed segments of the Hercynian Mass or severely deformed Mesozoic strata. The question of pedimentation of the extensive páramo limestone surfaces is not considered.

shales are too plastic and unstable to preserve planation "surfaces." It is even less reasonable to consider this area as an erosional surface when allowance is made for subsequent Pleistocene remodeling.

4) Although infrequent, some inaccurate citations can be found in Schwenzner's text. Most of these errors can, perhaps, be attributed to errors prevalent in the topographic information available at that time.

The confusion surrounding the time origins of the Pontian limestone páramo "C" surface (i.e., Pontian vs. pre-Pontian), introduced by Schwenzner, is perpetuated by Solé's (1952) treatment of the Meseta erosional surfaces. In reference to Pontian developments, Solé does not clearly distinguish between time-stratigraphic nomenclature and lithologic units. He refers, for example, to Miocene sedimentation and planation that was interrupted by post-Pontian deformation (Solé, 1952:172) in the one sense, and to a Pontian surface which is comprised of Pontian materials in the other.[8]

Solé's résumé of Schwenzner's contribution is appropriately thorough and a fair representation. Since Schwenzner's manuscript is not semantically clear in all places, and since its organization requires the reader to compile information from throughout the work in order to reconstruct a topical consideration, it is not unreasonable that misquotes occur. But Schwenzner (1937,71) holds the Rhodanian deformation to be post-Pontian, while Solé (1952,170) implies that Schwenzner held it to be Pontian. Other misquotes differ in kind but not degree.

The sequence of morphological evolution proposed by Solé (1952) and Birot and Solé (1954) would require that wherever vertically offset "C" and "D" surfaces are recognized, Tertiary fault lines also should exist to account for development of two erosional surfaces in a single planation period. This basic prerequisite would present a very complex fault system, particularly in the more central areas of the Tertiary basins, that has yet to be recognized in the study area.[9] Consequently, on the basis of this field work, their hypothesis is untenable for the Hesperian Meseta. It is possible, however, that post-Pontian tectonics did sufficiently disturb the base-level determinants of the Tajo Basin to effect a readjustment of planation processes at lower levels and perhaps account for the vertical break between the "C" and "D" surfaces, without recourse to a completely new planation cycle. Birot and Solé do not consider this alternative, however, and such an interpretation would not differ substantially from that of Schwenzner.

---

[8] " ... es indudable que por debajo de la superficie pontiense existe otro nivel ... " (Solé, 1952:215).

[9] Birot and Solé dealt for the most part with erosional surfaces offset by fault systems within the Cordillera Central. Establishment of the proper interrelationships among these several levels with those of the Meseta was not attempted.

Finally, the methods used and criteria followed by Birot and Solé (1954) in identifying planation surfaces are not explicit. Since the surfaces were not mapped in detail and since the Solé-Birot hypothesis required that any given surface need be allocated to only one of two planation cycles, these workers appear to have correlated surfaces on the basis of comparable or relative elevations. Such a procedure would obviously be of questionable value.

CHAPTER VI

PLEISTOCENE DEPOSITS OF THE ALTO HENARES

Introduction

The depositional record of a drainage basin preserves valuable evidence essential for a proper understanding of its geomorphologic history. Sediments are the aggradational counterparts to the denudational processes that sculpture erosional landforms. Sedimentological analysis and interpretation may well be the most singularly informative and reliable line of investigation, since an erosional landform provides the composite reflection of different processes and changing environmental conditions experienced throughout the dimension of time. In view of the deviations of Pleistocene climates from the "normal" conditions of geologic history, any attempt to unravel geomorphologic development without consideration of sediments and depositional landforms would be incomplete, if not misleading. A long succession of changing environments with new morphogenetic balances will lead either to a convergence of erosional forms or to the dominance of features which reflect one phase of particularly effective change. By contrast, the sedimentological record, if preserved, tends to provide unique data for the plurality of depositional environments which, in turn, further contribute to a definition of the regional environment and its respective place within the overall Pleistocene succession.

The following chapter provides a description and analysis of the sediments recorded in the Alto Henares. The nature of the various types of deposits is first discussed briefly to define the basic terminology. After noting the general distribution and stratigraphic relations of these deposits in an introductory fashion, the main part of the chapter presents a detailed description of the depositional suites for individual stream basins, based upon a selected number of key sites and exposures. Where appropriate, these materials are integrated into broader stratigraphic sequences and interpretations suggested for the different sediment categories.

Meseta today demonstrate that rainfall need not be plentiful. However, annual temperatures must be sufficiently high through most of the year to allow protracted and effective
evaporation of surface waters. The thick accumulations of buried marls in the study area
are essentially barren of macro and micro plant remains.

Conglomerates

Alluvial gravels derived from local bedrock comprise most of the clastic components of terrace conglomerates preserved in the study area. Quartzite gravels frequently
have a well-rounded shape but may also be fractured. Since intact Buntsandstein gravels
have similar properties, transport inferences based upon gravel morphometry are not justified. Limestone and dolomite gravels have frequently been rounded--at least in part--
by subaerial corrosion prior to erosion or subaqueous solution in the stream bed. As a
result, distance and turbulence of transport can seldom be assessed with confidence and
useful conclusions based upon gravel morphometry must be limited to quartzite gravels
that are not fractured or to limestone or dolomite gravels that have not been long exposed
to corrosion prior to transport. The gravels of the study area comprise fine-to-medium
grade pebbles cemented in light reddish-brown-to-red, calcareous matrices of coarse
sand. Bedding ranges from unstratified to well stratified and may include layers of well-
sorted materials. Cementation by diffuse calcite implies periods of inundation and impreg-
nation alternating with periods of desiccation. These conditions would be favored by
marked dry seasons with frequent periods of subsoil aridity. Whether such cementation
was post- or syn-depositional is normally difficult or impossible to determine since the
deposits are too shallow.

Slope Deposits

Breccia

Upland, midslope and footslope surfaces alike are littered with medium or coarse
size, limestone-derived rubble. In upland situations, the litter occurs as limestone regolith intermixed with thin soil horizons. Accumulations on midslopes and footslopes represent limestone scree derived directly from bedrock or, in part, from older colluvial
deposits. Rock fragments generally are corroded, suggesting protracted exposure to
chemical weathering and possible derivation from older material. In addition to the limestone fragments, breccias may include Keuper quartz crystals (known as Jacintos de
Compostela) or Buntsandstein (quartzite) materials reworked in a reddish-brown, coarse
sand matrix of soil sediments.

The clastic component of the cemented breccia is similar to modern detritus but
there is little or no contemporary production of limestone detritus by frost shattering in

the Hesperian Meseta (1000-1300 m), although frost weathering has been recorded in the Somosierra (1900 m) by Fränzle (1959, 57f). Production and supply of the breccia detritus relate, therefore, to colder or moister phases of the Pleistocene. This has been demonstrated in the case of the Costa Brava (Butzer, 1964a), for southeastern Spain (Wiche, 1961; 1964), and on Mallorca where breccias occur as lateral facies variants of limons rouges and as frost-weathered talus or scree (Butzer and Cuerda, 1962). In these subtropical lowland margins of the Mediterranean, the formation of breccia relates to moister early-glacial periods but in the Hesperian Meseta conditions were also much cooler. Production of detritus by intense frost shattering preceded calcite cementation, which may have been either synchronous with or subsequent to the intermixing of a soil mantle and regolith. Solifluction and frost transport also may have assumed roles in breccia formation, depending upon the severity of the climate.[1]

The breccias of the study area commonly preserve morphologically distinct remnant features that are peripheral to the present stream courses. Morphometry of the detrital components may be only subtly distinct from that of conglomerates and indicates they represent cemented surficial deposits that were marginal to ancient stream beds which deposited the conglomerates. These particular breccias differ from the masses of heterogeneous, cemented, boulder rubble described elsewhere in the Hesperian Meseta (Butzer, in preparation). The latter type represents a congeliturbate upland mantle, which may also have been characteristic of midslope and footslope zones; it is ascribed to the colder phases of the Pleistocene glacials.

Éboulis ordonnés

Accumulations of crudely stratified, angular screes are found in pockets on 23°-27° midslopes below the Pontian limestone. These fossil deposits consist of frost-shattered, fine-to-medium (2-10 cm) limestone fragments and may attain a thickness of 2 m. The facies are well consolidated but not cemented, with little or no interstitial material. Such detrital deposits are variously referred to as éboulis ordonnés or grèzes litées.

---

[1] The region south of the Cordillera Central in general is akin to the modern Mediterranean ecosystem rather than the decidedly more continental character of the northern part of the peninsula. However, on the Hesperian Meseta, Pleistocene conditions periodically were sufficiently less Mediterranean in nature to produce a periglacial type environment. This is indicated, for example, by relict periglacial features found 900 to 1000 m below the present lower solifluction limit of the Somosierra (Fränzle, 1959:34ff, 62). A periglacial depression of this magnitude would have placed the Hesperian Meseta well within the limits of Pleistocene cold climate phenomena, so that Pleistocene conditions would have been distinctive from those of the Mediterranean zone and perhaps also less rigorous than those of continental Europe. Consequently it must be recognized that Pleistocene morphogenetic analogies based upon subtropical, Mediterranean, or continental models may not pertain in the case of the Hesperian Meseta.

A different variety of this type of cold climate slope deposit occurs in certain foot-slope situations in the Mesozoic bedrock terrain. These beds comprise stratified, sub-angular, medium-grade detritus interlayered with stratified beds of finer detritus and coarse sand, locally indurated with calcium carbonate. The origin of this sorted type of éboulis ordonnés is thought to relate to seasonal downslope movement of surface wash over partially frozen ground (see Embleton and King, 1968:522f, with references).

Paleosols

Buried soils may serve as stratigraphic indicators of warmer interglacial phases within the broader Pleistocene context. Through an understanding of contemporary soil-forming processes and relevant climatic constraints, analogies may be drawn with regard to past pedogenetic settings.

The paleosols within the study area are analogous to the modern mantle of relict soils discussed in Chapter III. Differences are of degree rather than kind. So, for example, a relict terra fusca soil recorded in the Hesperian Meseta (Butzer, 1965) indicates an environment moister and perhaps warmer than the present environment, while local records of halomorphic paleosols recall the contemporary solonchaks of the campiña, differing perhaps only in depth and, therefore, duration of development.

The red Keuper clays lend problematic aspects to the interpretation of red paleosols and soil sediments since these are often quite similar to primary or derived Keuper parent material. Fresh exposures show that the plastic Keuper beds have undergone considerable internal kneading and dislocation and the structure may not be distinct from that of colluvial material derived from it. In addition, the Keuper grain-size spectrum may bear strong similarity with Pleistocene paleosols or soil sediments. Because of these analogies, recognition of red paleosols very often must depend upon careful evaluation of all data and paleogeographic aspects germane to a given situation.

Pleistocene Deposits of the Alto Henares

General Aspects

The Pleistocene deposits occur preeminently within the wide valley basins sculpted into the tableland páramos. In a few isolated cases breccias may be noted on these surfaces but for the most part deposits are embanked along the valley sides and bottoms. The oldest Pleistocene materials comprise the High Terrace, in the form of platform steps far removed from the present river courses, set along midslope segments but below the level of mesaforms or shoulders in the campiña that mark the youngest planation surface. Legacies of this terrace in the form of thick travertine accretions are found in each of the three Alto Henares basins studied.

Below this highest, ancient floodplain level but closer to the modern river channels are the better-documented fragments of the Middle Terrace. A variety of sediments and facies comprise the alluvial body so that it is the best known of the Alto Henares. Additional significance derives from the stratigraphic link provided by the Middle Terrace to adjacent drainage basins of the Hesperian Meseta.

Least obvious in terms of morphologic distinction are the Low Terrace alluvia of the upper Rio Henares, Rio Salado and Rio Dulce. These materials provide absolute dates for the Low Terrace and comparable sorts of information have been obtained from the flat expanse of the valley bottom that borders the major rivers; this is referred to as the Campiña Terrace.

With the exception of the campiña surface proper, characterized almost universally by huerta cultivation, the various segments of individual terrace levels occur discontinuously within the valleys so that terrace levels cannot be simply followed from point to point along the river valleys. In addition, with but one exception in the upper Henares, the full suite of Pleistocene terraces is nowhere preserved at a single location. The scattered distribution of these features requires, therefore, that the terrace remnants of the Alto Henares be interrelated on the basis of morphologic, altimetric, and sedimentologic properties supplemented in each case by careful field observations. The subsequent sections of this chapter outline the more salient aspects of these properties with an emphasis on the sedimentological record in the way of formulating a coherent framework for the fluvial terraces of the Alto Henares.

The Rio Henares System[2]

The Horna area

At Horna, the Arroyo de la Fuencilla and the Barranco de Valdemuco converge to form the Rio Henares. The arroyo is the major upper Henares channel draining the western flank of the Duero-Ebro-Tajo divide, which at this point is formed by the Sierra de Ministra. Horna is situated on a massive travertine platform breached by the arroyo (see Fig. 19). More than 2 m of well-cemented, cryptocrystalline, partially porous, travertine beds occur at the village (1090-1092 m) and atop the left-bank winnowing platform south of Horna. Morphologic remnants of this terrace reoccur for more than 1 km to the southwest.

A sequence of younger, freshwater deposits embanked against this travertine terrace is exposed just above the Arroyo de la Fuencilla channel. The basal strata rest on

---

[2] Terminology used in the description of Pleistocene materials is defined in Appendix A. Place names and locations used throughout this chapter relate to the latest available 1:50,000 editions of the Mapa Topográfico Nacional del Instituto Geográfico y Catastral.

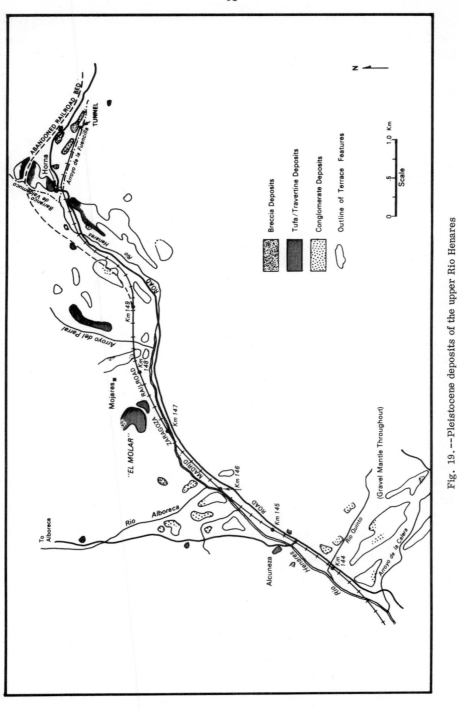

Fig. 19.--Pleistocene deposits of the upper Rio Henares

Keuper marl, as do the uppermost tufa deposits truncated by the highway. In all, 15 m of sediments are recorded by these two exposures between 1072 and 1087 m. They comprise, from base to top:

a) Over 20 cm of gray-to-dark gray, clayey silt (Table 4, No. H615). This consolidated, organic deposit has coarse, angular, platy structure and contains abundant shell fragments; more rare are small pockets of calcareous, medium coarse-to-coarse sands, and charcoal fragments. The body of sediment is laced with subhorizontal bands of oxidation. Clear, wavy contact.

b) 85 cm of loose, powdery marl, 97% HCl soluble. Color of the loose structured sediment varies from white (10YR 8/2) to pink (7.5YR 7/4) in wavy, stratified bands with some beds becoming more granular than others and with pockets of tufa detritus. The non-carbonate residual, which was too limited for grade-size analysis, comprises dark grayish-brown (10YR 4/2) to dark brown (7.5YR 4/2) clayey silt with fine sand. Diffuse, wavy contact.

c) 60 cm of light gray (10YR 7/2) tufa, 96.5% HCl soluble. The laminated beds contain shell fragments. The non-carbonate residue is a brown (7.5YR 5/2) clayey silt in negligible amounts for grade-size analysis.

The tufa (c) of this lower exposure at 1072 to 1074 m is re-exposed farther upslope, between 1079 and 1087 m by the road cut. As much as 8.7 m of subhorizontally bedded, organic tufas rest on Keuper material with an inclined discordance of 13°. Massive, almost pure very pale brown-to-yellow (10YR 7/3-8/8) tufas with some shell fragments are interlayered with well-cemented, brittle, tubular, white (2.5Y 9/2) organic tufas that contain occasional dolomite pebbles. The radiocarbon age of the upper massive tufa beds is 30,650 B.P. ±1350 (I-3111). This sequence of deposits is recorded 5 to 20 m above the former course of the Arroyo de la Fuencilla, which was diverted during railroad construction.

North of the village the abandoned railroad grade transects some 6 m of complexly interbedded clastic and precipitate deposits (Fig. 20). The upper strata are subhorizontal but the dip increases to 8° within the lowermost beds. The sediments are exposed between 1085 and 1095 m and are gently inclined in an overall southerly direction. From base to top:

a) Over 90 cm of white marl (Table 4, No. 1425). In part massive and in part thin-bedded, compaction of the sediment is variable, from unconsolidated in the basal sector that overlies Keuper bedrock, to broadly undulating, well-cemented sediment in the uppermost layers. Abrupt, wavy contact.

b) 95 cm of white marl (Table 4, No. 1426), distinctive on account of massive bedding and very weak, unconsolidated structure. This sediment thins horizontally, almost fingering out in places. Abrupt, smooth contact.

c) 55 cm of white, indurated marl (Table 4, No. 1427). Massive stratification of this deposit varies in structure from massive to laminated, to porous or even tufaceous. Abrupt, wavy contact; disconformity.

d) Approximately 20 cm of unconsolidated, white marl (Table 4, No. 1428). The massive bedding has very weak structure. Abrupt, wavy contact.

Fig. 20.--Setting of deposits north of Horna. More resistant conglomerates and indurated marl facies outcrop along the south slope above the abandoned railroad bed. The hill at the top comprises backfill.

e) 60 cm of gravels in reddish-yellow (7.5YR 6.5/6), sandy silt and granules. The crudely stratified, subrounded gravels of this semicemented conglomerate comprise carniolas limestone. The uppermost levels grade laterally into a very pale brown-to-light gray (10YR 8/3-6.5/2), indurated calcrete with some snail shell fragments, and medium grade gravel in a very stoney, marly silt. These deposits are in part disconformably overlain by Bed g, but also laterally grade into Bed f by an abrupt, wavy contact.

f) 40 cm of white, chalky marl (Table 4, No. 1430). The consolidated marl is massively bedded and bears traces of corrosion in the uppermost layers. Where laterally conformable with Bed g, overall thickness increases to 120 cm of reddish-yellow (5YR 5-6/6), semicemented, well-stratified conglomerate. The thin-bedded, medium grade, subangular gravels are in a matrix of medium-sandy silt or granule sand. Abrupt, wavy contact; disconformity.

g) 35 cm of variable marl (Table 4), ranging from crudely stratified, semiconsolidated, light yellowish-brown, partly oxidized, clayey silt or medium-sandy marl with blocky to platy structure (No. 1431a and b); to stratified or laminated, consolidated, light gray, locally oxidized marl (No. 1431c). Freshwater snails are locally abundant in the latter variant. Abrupt, wavy contact.

h) 30 cm of yellowish-red, clayey silt (Table 4, No. 1432). This crudely stratified, semiconsolidated sediment is of very coarse, angular, blocky structure and contains local gritty lenses of semicemented, reddish-yellow (7.5YR 6/6), granule matrix conglomerate with fine-to-medium, subangular pebbles and croûte zonaire laminae. This bed, as well as Bed g, locally, grade laterally into approximately 55 cm of yellowish-red (5YR 5/6), crudely stratified, semiconsolidated, clayey silt that is similar to Bed g, but distinctive because of fine, evaporite accumulations along the bedding planes of ped faces and because of subhorizontal lenses of incipient calcretion. Abrupt, wavy contact; disconformity.

i) 15-40 cm of medium grade gravel in a pink (7.5YR 8/6-5YR 8/4), sandy silt. The subrounded, limestone gravels are well stratified in a semicemented, calcrete matrix. This sediment is laterally interfingered with lenses of partly concreted, gravelly soil wash. Clear, wavy contact.

j) 30 cm of reddish-yellow, medium-sandy marl (Table 4, No. 1436). The stratified, semiconsolidated sediment contains subrounded, medium grade gravel within a medium, subangular blocky structure. Local consolidation records zones of incipient calcareous enrichment. Clear, wavy contact.

k) 50 cm of very pale brown, subrounded gravel in a very pale brown (10YR 7/4), sandy silt. The fine-to-coarse grade gravels of the semicemented, conglomerate are well stratified and contain local calcareous concretions. Clear, wavy contact; disconformity.

l) 60 cm of very pale brown marl (Table 4, No. 1438). This well-stratified to laminated, semicemented, subhorizontal deposit fills shallow draws in Bed k, and contains calcreted, subangular to subrounded, gravelly soil wash with some limited surface corrosion. Beds i through l thicken to the northeast segment of the section, attaining an overall thickness of 2.6 m.

This complex sequence of deposits records changing sedimentary situations in the headwater region of the ancient Rio Henares (see Fig. 21). The marly units indicate floodplain-type backswamps or ponded lacustrine settings of a local nature. The variable induration of the several, distinct, basal marl facies very likely is indicative of overall wetter or drier conditions. Accumulation of marls in the upper half of the sequence was contemporaneous with the aggradation of fluvial-type gravels and/or screes of lateral fans, which ultimately were supplanted by soil wash. The overall sequence represents the interplay of lacustrine, floodplain, and colluvial deposits in a situation that was marginal to an ancient river channel.

Several distinct periods of alluviation are indicated by the deposits in the vicinity of Horna. The extensive travertine platform at the village and south of it record the oldest phase. Following partial destruction of this platform, the marl and conglomerate sequence north of Horna was graded to a valley bottom or floodplain altimetrically lower than the travertine platform (see Fig. 21). The organic tufa and marl deposits immediately south of Horna record a third and still younger alluviation phase which was adjusted

95

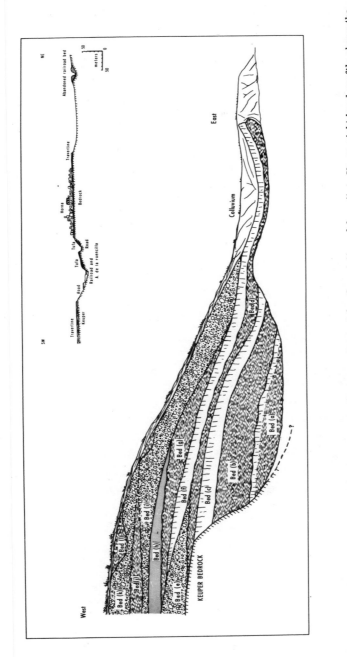

Fig. 21.—Composite section of deposits north of Horna. See text for description of deposits. Upper right hand profile shows the relationship of this exposure to the tufa and travertine deposits at Horna.

to a floodplain 5 to 20 m higher than the general level of the modern valley, or campiña, throughout this area. The sediment characteristics of these deposits are listed in Table 4.

TABLE 4

SEDIMENT CHARACTERISTICS IN THE HORNA AREA

Sample Number	Strata	Color	Texture	Sorting	CaCO$_3$ % wgt	pH
			Low Terrace			
H615	Bed a	10YR 5/1-4/1	Clayey silt	Moderate	24.2	7.3
			Middle Terrace			
1425	Bed a	10YR 8/2	Marl	–	99.5	8.4
1426	Bed b	10YR 8/2	Marl	–	74.8	7.4
1427	Bed c	10YR 8/2	Marl	–	99.4	8.4
1428	Bed d	2.5Y 8/2	Marl	–	61.5	8.2
1430	Bed f	2.5Y 9/2	Marl	–	98.2	8.3
1431a	Bed g	10YR 6/4	Clayey silt	Moderate	29.8	7.6
1431b	Bed g	10YR 6/4	Medium-sandy marl	Moderate	37.8	7.7
1431c	Bed g	2.5Y 7/2	Marl	–	82.4	7.9
1432	Bed h	5YR 5.5/6	Clayey silt	Moderate	25.7	7.6
1436	Bed j	7.5YR 6/6	Medium-sandy marl	Poor	62.0	8.0
1438	Bed l	10YR 8/3	Marl	–	87.2	8.3

Additional exposures of Pleistocene materials, both upstream and downstream from Horna, record each of the three separate periods of alluviation noted above. The exposures comprise travertine, breccia and conglomerate deposits scattered along both banks of the Henares (see Fig. 19). Along the right bank, for example, between Horna and Mojares, deposits cap smoothly rounded interfluve spurs with similar morphology and elevation within the valley bottom. Viewed in the field, their accordance defines a single, continuous surface but in reality fluvial dissection has effected local relief in the order of 30 m. Conglomerate and travertine deposits that mantle adjoining spurs demonstrate horizontal facies variation of the same alluviation cycle, and recall the conglomerate-marl lateral facies change north of Horna. As much as 3 m of dense, well-cemented, pale brown (10YR 6/3), cryptocrystalline travertine interlayered with very pale brown (10YR 8/3), dense porous tufa-like travertine containing shell fragments cap such a spur (1064-1070 m) north of Km 149 of the Madrid-Zaragoza railroad (see Fig. 19). Subangular dolomite and limestone gravels are imbedded in the basal portion of the travertine, which overlies a thin (11 cm) bed of gray-brown (10YR 5/2), coarse-sandy marl that grades downward into yellowish-red (5YR 5/6), sandy, gravel sediment. The gravelly basal portions of this travertine cap and the underlying marl and sediment provide additional indications of the lateral conformability of the clastic and precipitate facies north of Horna that are described above.

Deposits near Mojares

Still another facies variation of the Horna conglomerate-marl sequence occurs farther downstream along the flanks of the valley spur west of the Arroyo del Parral (Fig. 19). Here the railroad cut has exposed more than 3 m of clastic deposits dipping 34° to the west between 1036 and 1040 m. The contact is obscured, but at least 75 cm of the basal sediments of the sequence overlie Keuper marl and comprise very pale brown (10YR 7/3), silty sand containing occasional subrounded gravels, pale olive (5Y 6/3) clay lenses, oxidation nodes and many small shell fragments. These sands are overlain by 1.8 m of fine-to-medium, subrounded or subangular gravel, varying from crude to well stratified in the uppermost layers. The deposits are weakly indurated and interspersed with wavy, calcareous bands; the lower bands comprise very pale brown (10YR 7/3), 10 to 12 cm laminated crusts with subhorizontal oxidation streaks. The uppermost calcareous layers vary from thin, locally oxidized, clay or silt lenses, to calcrete 1 to 4 cm thick. The topmost sediments of the exposure comprise 1.3 m of well-stratified, subangular-subrounded gravels. The gravels are in a light gray-to-light brownish-gray (10YR 7/2-6/2), calcite-cemented, medium coarse sand or granule matrix containing fragments of freshwater snail shell. Unstudied pieces of corroded mammalian bone and teeth, probably _Equus_ sp., were recovered from the uppermost layers of these deposits.[3]

Additional exposures of these sands and gravels occur along the same valley spur above the Arroyo del Parral. The sediments contrast with the gravel facies described north of Horna primarily on account of the steeply inclined bedding, weaker induration or cementation, and more sandy matrix. The silt and clay lenses, snail fragments, oxidation strains, and worn edges of crystalline quartz granules (Jacintos de Compostela) demonstrate fluvial origins for these deposits. Gradual contacts and internal conformity occur throughout the deposits, so that the inclination may be the result of undermining and slope failure when the deposits were in a thoroughly frozen condition. Additional evidence of this type of occurrence, discussed in a later section of this chapter, is found elsewhere in the Alto Henares.

---

[3]An indication of solution rates under present conditions is provided by the complete weathering-out of the bone fragments during the 18 months that intervened between field seasons. Initially the bone was firmly imbedded in a small crevasse of the cemented conglomerate. Carbonation resulting from surface runoff freed the bone and washed it downslope in the 1.5 year interval.

The most striking Pleistocene morphologic feature of the upper Henares is the imposing, flat-topped platform southwest of the village of Mojares known as "El Molar" (Fig. 22).[4] The surface of this terrace (1075-1090 m) comprises approximately 12 m of dense, cryptocrystalline, laminated travertine gently inclined 5° toward the present course of the Rio Henares. The travertine rests on reddish or yellowish-brown (7.5YR 4/2-3.5YR 5/4), sandy, marl-like sediment with subrounded limestone or dolomite pebbles. Locally, the basal travertine phases contain gravel facies, such as observed in a cave-shelter (1045 m) at the southwest base of the platform where subrounded-subangular, corroded limestone, dolomite and travertine pebbles are intermixed within the body of the travertine beds.

Less than 15 cm of a dark brown (7.5YR 3/2), loose, sandy rendzina A-horizon mantles most of the travertine surface along with a large amount of scattered travertine detritus. At the northeast margin of the platform, the surface litter includes blocks of well-cemented breccia, up to 1.3 m in diameter. The light, reddish-brown (5YR 6/4) breccia comprises fine, medium and occasional coarse size subangular limestone, dolomite, and quartzite detritus in calcite-cemented medium sand, fossilized, surficial wash. This particular type of slope breccia is well recorded throughout the Alto Henares. It is preserved elsewhere in the upper Henares upstream from Horna, at the Henares-Alboreca confluence, near Alcuneza, and above the Arroyo de la Calera (see Fig. 19).

The arrangement of intact travertine blocks along the west and east flanks of El Molar reflects the step-like profile of the south face (Fig. 23). The east flank is rimmed by an intact travertine ledge inclined 7° toward the Rio Henares. The uppermost extremities of this ledge merge with the top surface but downslope a broad, 50 m platform emerges at 1052 to 1063 m on the south face to establish a conspicuous bench consisting of more than 5 m of massive travertine. A lower, narrower, 1041 to 1046 m bench of 5.5 m of massive travertine with basal gravelly facies rests on Keuper bedrock. The travertine-Keuper contact facies and the esplanade aspect of the El Molar profile demonstrate the origin of this feature. Precipitation of the travertine materials occurred in riverine situations, as shown by the basal gravel or marl facies, or in marginal floodplain locations as indicated by the underlying surficial wash. The great thickness of the massive layers indicates extended periods of aggrading by carbonate-charged waters. The initial phase of accumulation is recorded by the uppermost platform surface. Partial destruction of this preceded aggrading of the middle, 1052 to 1063 m platform. Reinitiation of this cycle led to formation of the lowermost bench which subsequently was dissected by the present course of the Henares channel. The polygenetic development of El

---

[4] This feature, so named by the villagers, is not to be confused with the interfluve spur at Km 149 of the railroad referred to as "El Molar" by Zaranza (1964).

Fig. 22.--El Molar. Two distinct platforms evident along the south face profile at the left comprise this massive travertine feature. The Rio Henares is in the foreground.

Molar is corroborated by the occurrence of several terrace levels throughout the upper Henares and by other travertine benches elsewhere in the Alto Henares system.

The Alcuneza area

Pleistocene deposits in the vicinity of Alcuneza include travertine and tufa precipitates and fluvial conglomerates. Travertine materials occur above both banks of the Rio Alboreca, embanked against Keuper midslopes, and at Alcuneza where more than 2 m of dense, subhorizontal, travertine forms a platform at the village that is offset 30 m above the modern Henares channel. On the opposite bank of the river, as much as 3 m of lower

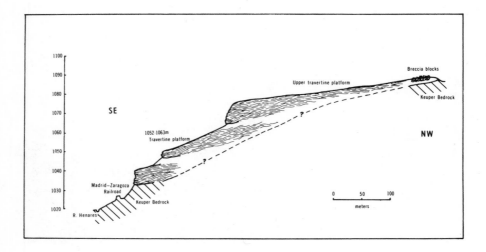

Fig. 23.--Transect at El Molar

(1013 m) and younger tufa deposits are preserved in a situation that was marginal to the river that cut into and partially destroyed the Alcuneza travertine terrace.

Two distinct levels of conglomerate deposits occur along the Rio Alboreca. Three remnants of a high platform are found above the right bank between 1040 and 1046 m. These same conglomerate gravels also are interbedded with the uppermost layers of the Alcuneza travertine accumulation. A lower (1025 m) left bank conglomerate terrace is recorded at the Alboreca-Henares confluence (Fig. 24). Morphologic remnants, without deposits, of this same terrace are exceptionally well preserved above the right bank of the confluence (see Fig. 19 for the location of deposits mentioned in this section). A

Fig. 24.--Low Terrace conglomerates near Alcuneza. The conglomerates rest on Keuper bedrock. The Rio Henares occupies the valley floor at the right.

panorama of the Alboreca-Henares confluence is shown in Figure 25. The macro descriptions of the deposits mentioned above and shown in this photograph are presented in Table 5.

The Arroyo de la Calera

Fluvial gravels and conglomerates mantle much of the valley bottom along the Rio Henares left bank, between Alcuneza and the mesaform residual northeast of Sigüenza (Fig. 26). The quartzite gravels were deposited by streams that originated in the Buntsandstein core of the Sigüenza anticline, breached the Muschelkalk hogback, and spewed

TABLE 5

MACRO CHARACTERISTICS OF SOME CEMENTED DEPOSITS NEAR HORNA AND ALCUNEZA

Sample Number	Location	Cementation	Stratification	Clastic Size	Color	Matrix Texture
Conglomerates						
H223	Arroyo de la Calera, Middle Terrace	Weakly cemented	Unstratified	Medium	Pink (7.5YR 7/4)	Medium-to-coarse sand
H224	Arroyo de la Calera, Middle Terrace	Cemented	Crude	Medium-Coarse	Light reddish-brown (5YR 6/3)	Medium-to-coarse sand
H508	Km 149, Middle Terrace	Cemented	Unstratified	Fine-Medium	Brownish-yellow (10YR 6/8)	Fine-to-coarse sand
H606	Below Horna, Middle Terrace	Cemented	Crude	Medium	Very pale brown (10YR 7/4)	Coarse sand
H607	Km 150, Middle Terrace	Cemented	Crude	Medium	Pink (7.5YR 7/4)	Coarse sand
H706	Henares-Alboreca confluence, Low Terrace	Cemented	Unstratified	Fine-Medium	Yellowish-brown (10YR 6/4)	Coarse sand
H707	Alcuneza, Middle Terrace	Well-cemented	Crude	Medium-Coarse	Pale brown (10YR 6/3)	Medium-to-coarse sand
Breccias						
H105	El Molar, Cap	Cemented	Unstratified	Detritus	Pink (7.5YR 7/4)	Fine sand
H221	Arroyo de la Calera, Cap	Well-cemented	Unstratified	Grit-detritus	Brownish-yellow (10YR 6/6)	Medium sand

103

Travertines

Sample Number	Location	Color	Porosity	Carbonate %	Other Properties
H601	Above the Arroyo de la Fuencilla	White (10YR 8/2)	Porous	99.5	Pale brown (10YR 6/3) calcite veins
H505	Km 149, Middle Terrace	Pale brown (10YR 6/3)	Dense	98.4	Rare fine gravel
H703	Rio Alboreca, Middle Terrace	Light reddish-brown (5YR 6/3)	Porous	93.9	Rare fine sands
H704	Alcuneza, Middle Terrace	White (10YR 8/2)	Dense	99.5	Recrystallized

104

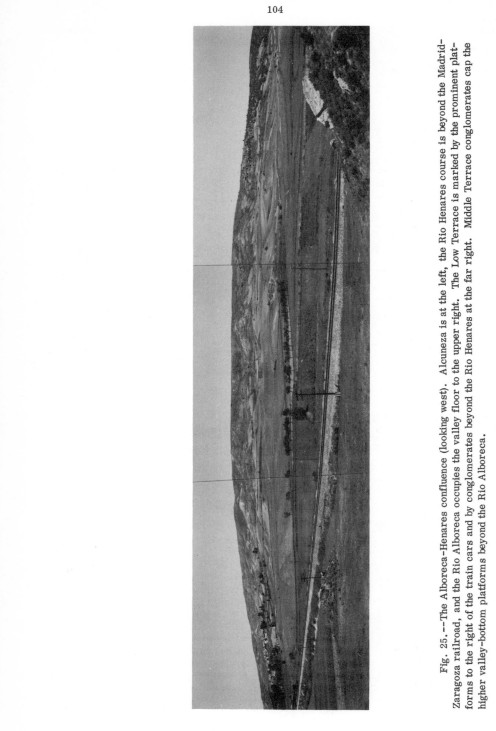

Fig. 25.--The Alboreca-Henares confluence (looking west). Alcuneza is at the left, the Rio Henares course is beyond the Madrid-Zaragoza railroad, and the Rio Alboreca occupies the valley floor to the upper right. The Low Terrace is marked by the prominent platforms to the right of the train cars and by conglomerates beyond the Rio Henares at the far right. Middle Terrace conglomerates cap the higher valley-bottom platforms beyond the Rio Alboreca.

huge quantities of sand and gravel onto the Rio Henares floodplain. Later streams incised new courses and thereby established the older channel beds as terraces; the Rio Quinto and Arroyo de la Calera are the modern counterparts of streams that alluviated these terraces. The suite of platforms preserved in this segment of the valley is unique; nowhere else in the entire Alto Henares system does a succession of Pleistocene terraces occur in juxtaposition with diagnostic deposits (Fig. 27).

The highest fluvial platform slopes 2° toward the Rio Henares between 1056 and 1064 m. The surface is densely littered with fractured, subrounded, medium quartzite gravels (Table 15) interspersed within as much as 3 m of loose, brownish-yellow-to-yellowish-brown (10YR 6/5-5/3.5), medium-to-medium coarse, sand and granule alluvium. Occasional small blocks of unsorted, unstratified breccia also occur on this platform. The well-cemented, brownish-yellow (10YR 6/6), medium sand breccia matrix binds quartzite and limestone detrital materials.

The more extensive middle platform is characterized by gravel and conglomerate deposits. This terrace exhibits varied morphologic expression because of later dissection, and since gravel mantles cover all of the fluvial platforms of the Arroyo de la Calera, elevation of the Middle Terrace is best identified by the conglomerates. Outcrops of this crudely stratified, medium-to-coarse, quartzite and limestone conglomerate in a pink (7.5YR 7/4), sandy matrix are exposed between 1026 and 1040 m along the sides of the terrace. The body of the terrace alluvium comprises more than 10 m of red (2.5YR 4/6), reddish-yellow (5YR 6/6) or brown (7.5YR 5/4), medium-to-medium coarse sand and granules. Fractured, rounded, medium quartzite and limestone gravels (Table 15) are interspersed throughout the alluvium.

Massive outcrops of the Middle Terrace conglomerate also are exposed above the Rio Quinto right bank at 1033 and 1035 m (see Fig. 28). At least 6.3 m of the deposits are evident but the overall accumulation may be as much as 10.6 m. This cemented, well-stratified conglomerate has sorted bedding; coarse, subangular-to-subrounded limestone gravel in a pink (5YR 7/4), coarse sand matrix 40 to 55 cm thick are interbedded with 15 to 50 cm stratified, subrounded pebbles cemented in a reddish-yellow (5YR 6/6) coarse sand matrix. The distinctive bedding of these Middle Terrace conglomerates records the interlayering of Rio Quinto fluvial gravels and more angular, limestone detrital screes. The deposits accumulated in a footslope situation marginal to the river floodplain.

The surface of the lowest platform of the arroyo suite is inclined to the modern campiña of the Rio Henares. This surface is mantled by rounded but fractured, medium, quartzite gravel (Table 15). Lateral gullying along the flanks of the upper reaches of this terrace has exposed the component sediments (from base to top):

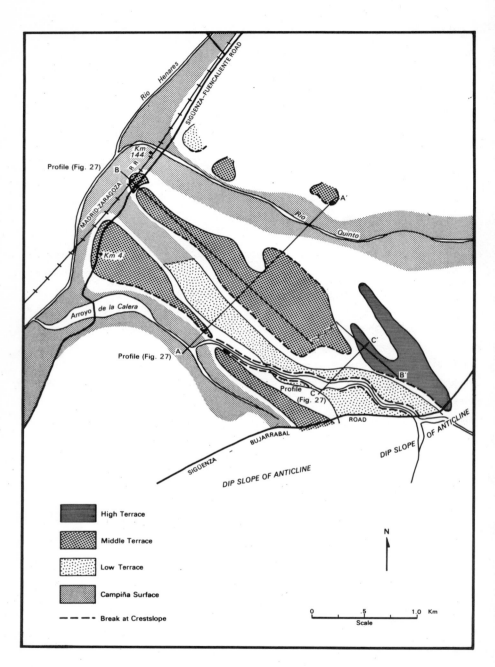

Fig. 26.--Arroyo de la Calera. The transects along lines A-A´, B-B´, and C-C´ are shown in Figure 27.

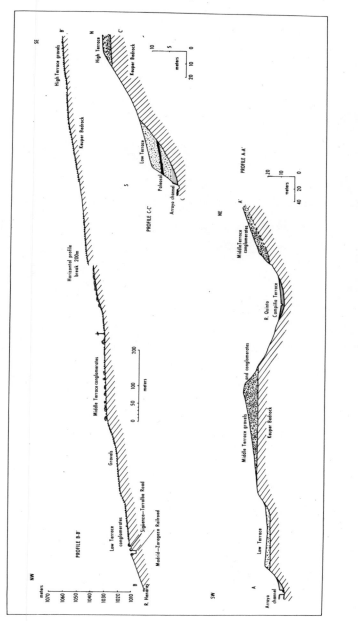

Fig. 27. --Transects at the Arroyo de la Calera. Location of the transects is shown on Figure 26.

Fig. 28.--Rio Quinto conglomerate deposits. More than 6.3 m of conglomerates are recorded at this location above the Rio Quinto right bank.

a) Over 220 cm of gray, silty clay (Table 6, No. H211). This consolidated, coarse blocky or prismatic alluvium has scattered pebbles and cobbles throughout, up to 20 cm in diameter in the basal portions where the deposits are laterally conformable with the modern arroyo channel gravels. The body of the sediment contains small (5 mm) calcareous grit and occasional charcoal fragments. Clear, irregular contact.

b) 45 cm of reddish-brown, clayey sand (Table 6, No. H210). The sand and larger fractions are generally dispersed throughout the consolidated clayey silt-to-fine sand matrix. Traces of charcoal occur within the sediment. Gradual, irregular contact.

c) 35 cm of brown, silty sand (Table 6, No. H209). This consolidated horizon has platy structure with local networks of micro calcareous bands and root casts. Color contrasts mark this humic, paleosol horizon. Clear, wavy contact.

TABLE 6

SEDIMENT CHARACTERISTICS OF THE LOW TERRACE IN
THE ARROYO DE LA CALERA AND AT BAIDES

Sample Number	Strata	Color	Texture	Sorting	Ca Co$_3$ % wgt	pH
			Arroyo de la Calera			
H211	Bed a	5Y 6/1	Silty clay	Good	28.2	7.1
H210	Bed b	5YR 4.5/3	Clayey sand	Moderate	2.6	7.1
H209	Bed c	7.5YR 5.5/2	Silty sand	Moderate	11.4	7.2
H208	Bed d	10YR 6/3	Sandy clay	Poor	12.7	7.3
			Rio Henares at Baides			
H316	Bed a	10YR 7/3	Clayey coarse sand	Poor	44.8	8.0
H315	Bed b	10YR 6/2	Clayey coarse sand	Moderate	27.1	8.1
H314	Bed c	10YR 7/3	Sandy marl	Poor	55.1	7.6
H306	Bed d	10YR 6.5/3	Coarse sand	Poor	34.2	7.6
H305	Bed e	5YR 6.5/1	Silty fine sand	Moderate	59.4	7.6
H313	Bed fi	10YR-2.5Y 7/2	Marl	Moderate	47.5	7.7
H312	Bed fii	10YR 7/1	Clayey medium sand	Moderate	41.3	8.0
H311	Bed fiii	10YR 7/1	Marl	Good	54.1	7.5
H310	Bed g	10YR 6/2	Marl	Moderate	78.3	7.8
H309	Bed g	10YR 7/1	Marl	Moderate	76.0	8.0
H308	Bed g	10YR 7/1	Marl	Moderate	73.2	7.6
H302	Bed h	10YR 8/2	Marl	-	98.7	7.9

d) 465 cm of pale brown, sandy clay (Table 6, No. H208). The consolidated, fine-to-medium, crumb-structured sediment contains occasional pebbles and comprises a colluvial overburden on top of the paleosol, derived from the adjoining higher terrace level.

The paleosol horizon is 3 m above the present arroyo floor and is graded toward the Rio Henares at 3° to 4°. The colluvial overburden thins in the same direction and both levels ultimately disappear before the present arroyo course breaches the Middle Terrace farther downstream. Termination of the organic horizon, which resembles a braunlehm vega (Kubiena, 1953:140f), can be estimated from aerial photographs by a gentle change in the longitudinal profile of the Low Terrace surface and the coincident darkening of surface material (see Fig. 26). Extension of the paleosol longitudinal profile toward the Rio Henares indicates that it was graded to a floodplain in the order of 10 m higher than the contemporary campiña. Such a floodplain is suggested further by conglomerate deposits at 1009 m partially buried at the foot of the Middle Terrace (Fig. 27), and at 1010 m above the Rio Quinto right bank near its confluence with the Rio Henares. These conglomerate materials are more than 15 m below the Middle Terrace conglomerates and represent a later aggradational phase.

The modern arroyo channel cut into the Low Terrace is graded to the Rio Henares course while valley-bottom surfaces tributary to the Henares are graded to the campiña surface (see Fig. 29). The truncation of the Low Terrace paleosol noted above most likely occurred with the regrading of these entrant surfaces to the modern campiña. The latter has been extensively reworked by huerta-type agriculture but quartzite gravels which locally mantle segments of this surface record the youngest fluvial surface referred to as the Campiña Terrace.

Fig. 29.--Arroyo de la Calera terraces. The modern arroyo channel is cut through Low Terrace deposits and into Keuper bedrock; farther downstream the Low Terrace surface merges with the Rio Henares campiña surface. The Middle Terrace defines the horizon. View is toward the Muschelkalk dipslope of the Sigüenza anticline.

The terrace suite preserved along the Arroyo de la Calera records the several Pleistocene fluvial cycles that are documented throughout the Alto Henares. While other individual exposures may relate the stratigraphic details, this sequence presents the broad framework. In general outline form, the sequence may be outlined as follows:

a) Planation of the High Terrace and alluviation of sandy gravels.

b) Dissection forming the Middle Terrace and deposition of conglomerates and gravels. Breccia formation of detrital accumulations on the High Terrace.

c) Protracted dissection of the Middle Terrace with temporary periods of equilibrium sufficient to allow lateral dissection, as indicated by several benches occurring in the Middle Terrace profile. Ultimately, alluviation of Low Terrace materials.

d) Morphostatic phase with soil development.

e) Fluvial planation of the Low Terrace and modern campiña alluviation.

f) Dissection (1–3 m) by the modern drainage network.

The Sigüenza area and the Henares gorge

Localized thin gravel mantles and fluvial conglomerates in the vicinity of Sigüenza represent downstream counterparts of Arroyo de la Calera terrace levels. These deposits have been preserved above the river right bank in situations that correspond altimetrically to the Low and Middle Terrace platforms at the arroyo. The occurrences are as follows: 1) weathered conglomerate 8 to 10 m above the Rio Henares opposite the La Obra school, at approximately 1007 m, 2) partially buried conglomerate at 995 m, 8 m above the river along a spur cut by the Madrid–Zaragoza railroad at Km 140.7, 3) rounded but fractured quartzite gravels 23 m above the Rio Henares, mantling a 1008 m spur immediately upstream from the Sigüenza railroad grade crossing (Table 15), and 4) large conglomerate blocks 25 m above the river on a platform adjacent to the Guardia Civil barracks.

Below Sigüenza the campiña surface (Campiña Terrace) narrows as the river course meanders through the gorge it has incised into the Mesozoic uplands. Where the valley bottom is wide enough, this surface comprises a cultivated flat 2 to 3 m above the river level. Downstream from Moratilla de Henares, the campiña broadens up to 400 m so that characteristic Pleistocene deposits once again occur within the valley floor margins in the form of dense, laminated, travertine accumulations between 981 and 990 m. These deposits outcrop above the left bank of the major arroyo valley southwest of Moratilla.

Farther downstream, near Estación de Cutamilla, the morphologic equivalent of the Moratilla campiña, to which defunct lateral drainage courses are graded, occurs 6 to 10 m above the river floodplain. The Rio Henares gradient increases rapidly at the Mesozoic-Tertiary bedrock contact near Estación de Cutamilla, so that the upper Henares campiña surface occurs here as a 6 m terrace that downstream increases in elevation above the modern floodplain. The deposits of this terrace are first exposed at Km 128.8 of the Madrid-Zaragoza railroad and additional sites are found all the way to Baides, at railroad cuts and river meander bends. The following exposure subsequently will be referred to as a Low Terrace type-site; it occurs 2 km upstream from Baides above the Henares left bank, near the Arroyo de Barrero entrance. From bottom to top (Fig. 30):

a) Over 20 cm of very pale brown, clayey coarse sand (Table 6, No. H316). The loose, sandy sediment contains abundant subrounded, fine-to-medium grade limestone pebbles. Gradual, wavy contact.

b) 10-15 cm medium-to-coarse pebbles, in a light brownish-gray, clayey coarse sand matrix (Table 6, No. H315). The subrounded or rounded gravels are dispersed throughout this loose, humic horizon. Clear, wavy contact.

c) 12-15 cm of very pale brown, sandy marl (Table 6, No. H314). The weakly consolidated, prismatic structure contains abundant small charcoal fragments and chips, as well as occasional small shell fragments. Local traces of ferruginous staining indicate oxidation. Abrupt, wavy contact.

d) 360 cm of subrounded, medium, quartzite, limestone and dolomite gravels (Table 15) in a pale brown, coarse sand matrix (Table 6, No. H306). Gravels are well stratified with a tendency for sorting into well-stratified, 10 to 20 cm thick layers or bands of fine-to-medium and medium-to-coarse size pebbles. The texture of the body of the sediment varies considerably, vertically and laterally, from sandy marl to the sandy gravel bands, but without disconformities. Local pockets of oxidation are recorded by ferruginous stains. The topmost sandy marl sediments develop a fine prismatic structure. Abrupt, wavy contact; disconformity.

e) 15 cm of light gray, pebbly, silty fine sand (Table 6, No. H305). The topmost 10 cm have a fine prismatic structure with increased organic content. Small shell fragments and calcareous grit contribute to the generally high carbonate aspect of the organic, floodplain-type sediment. Clear, wavy contact.

f) 65 cm of interlayered marls and sands, from base to top:

i) 20 cm of light gray marl (Table 6, No. H313). The lower half is weakly consolidated and unstructured, but the uppermost 10 cm show fine prismatic structure with shell and charcoal fragments and oxidation stains, perhaps along rootlet canals. Abrupt, plane contact.

ii) 15 cm of light gray, clayey medium sand (Table 6, No. H312). The weakly consolidated, crudely stratified sand of coarse, columnar structure is rich in sand-size calcareous shell and shows some evidence of oxidation. Radiocarbon date, 19,450 B.P. ±350 (I-3545). Abrupt, plane contact.

iii) 30 cm of light gray marl (Table 6, No. H311). The lower 15 cm are similar to (ii) above. The topmost 15 cm of laminated, weakly consolidated marl are of fine prismatic structure containing shell and charcoal fragments. The upper 2-5 cm become organic. Clear, wavy contact; disconformity.

g) 70 cm of light-to-brownish-gray marl (Table 6, No. H310, H309, H308). The body of the sediment is weakly consolidated and massively bedded with tufa splinters and debris strongly concentrated in the middle zone of the sequence. Small shell fragments are prominent in the upper 15 cm but also occur throughout, and small charcoal flecks and oxidation stains are found locally. The deposits contain a sizeable calcareous sand fraction (not evident in the tabular data because of HCl treatment). Clear, wavy contact; disconfromity.

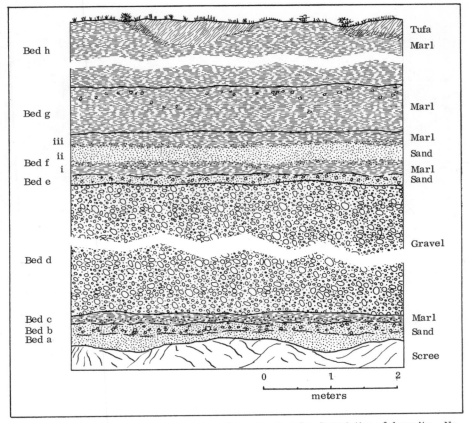

Fig. 30.--Baides Low Terrace section. See text for description of deposits. No vertical exaggeration.

h) Over 250 cm of white marl and tufa (Table 6, No. H302). The massively bedded marl and tufa materials include well-stratified horizontal and inclined lenses with reworked organic tufa, or slightly organic sediments. Radiocarbon date, 6560 B. P. ±130 (I-3110).

The deposits comprise a terrace 14 m above the river channel at 895 m.[5] The basal sediments of the sequence are offset 3.5 m above the campiña floor, into which the Henares is incised some 3 m (Fig. 31). The lateral constriction of the gorge has prevented

---

[5]Facies variations occur prominently within the Low Terrace in the immediate vicinity of the type-site. About 100 m downstream, an adjoining Low Terrace spur comprises more than 5 m of interdigited gravel, marl and silt strata, also with several organic horizons but showing lateral facies variation from the type-site. Across the river, more than 9 m of dense, stratified tufa deposits record a further variation of Low Terrace deposition.

Fig. 31.--Baides Low Terrace. View to the east with the Rio Henares at the left of the photograph downcut into the campiña surface.

development of a campiña surface in the same sense as in the upper Henares valley. The sequence may be summarized as follows:

a) Floodplain alluviation of sands with fine gravels.

b) Morphostatic phase with soil development of <u>borovina</u> type, followed by marginal floodplain alluviation of sandy marl with periodic inundation.

c) Protracted alluviation of gravels and sands with periodic inundation.

d) Erosional disconformity.

e) Floodplain alluviation.

f) Second phase of floodplain soil development (<u>borovina</u>).

g) Renewed alluviation of floodplain sediments yielding a Late Pleistocene radiocarbon date.

h) Erosional disconformity.

i) Floodplain sedimentation.

j) Erosional disconformity.

k) Protracted ponding of carbonate-charged waters and the precipitation of tufas and marls; Holocene radiocarbon date.

The isotopic dates demonstrate that aggrading of the Baides Low Terrace continued through the Late Pleistocene and Holocene, interrupted by several periods of erosion. These dates, and that from Horna, show that upstream of the Mesozoic-Tertiary contact in the Henares gorge, the Late Pleistocene Low Terrace and Holocene Campiña Terrace enjoy separate and distinct morphologic and altimetric expression; on the other hand,

downstream from this point, the counterparts of these levels are embodied in a single,
Low Terrace. The exact relationship of the several disconformities at the Baides site to
the geomorphic development of the upper Henares valley is uncertain.

Synthesis

Fluvial deposits occur above the Rio Henares campiña in association with three terrace levels. The distinctiveness of these phases of alluviation is demonstrated in the vicinity of the Arroyo de la Calera where individual levels with deposits are offset one from
another. The locations of the Pleistocene deposits in the upper Henares are shown in Figure 19, while the stratigraphic interrelationship of these terrace fragments, as outlined
below, is illustrated by the longitudinal profiles of Figure 32.

The 40 to 45 m High Terrace[6]

The oldest alluviation phase is recorded by travertine and gravel deposits. Platforms of the former are preserved at Horna 28 to 32 m above the river channel and east of
that village (+52 m), and again southwest of Mojares at El Molar (+55-70 m).[7] The High
Terrace is recorded at the Arroyo de la Calera by a quartzite gravel mantle (+62 m). The
absence of travertine accumulations among the arroyo deposits may be explained by the
fact that throughout the Alto Henares these deposits only occur below the Jurassic or Cretaceous limestone uplands and not in association with the Muschelkalk limestone. The
High Terrace marks a general floodplain stage in the upper Henares basin 40 to 45 m
above the modern campiña.

---

[6]Reference to terrace levels (elevations) in relation to the ancient floodplain that
they record is a common practice. Reconstruction of these levels using existing terrace
fragments allows standardization in terms of a single elevation or elevation range. This
procedure is adopted here but the large range of generalized altimetric values of a particular terrace remnant among the several basins deserves comment. Among the factors
involved is the fact that laterally shifting channels in the easily eroded Keuper have spread
deposits over a wide area such that modern river channels, to which terrace elevations
are referred, may be considerably distant from and below ancient floodplains thereby intro-
ducing differing elevations for an individual terrace level. In addition, the elevation of
specific terrace level above modern river courses tends to increase downstream as a consequence of the steepening of longitudinal profiles. Because of these factors, and perhaps
additional local ones, relative elevations of particular terrace fragments are widely dis-
parate in the Alto Henares.

[7]Relative elevations stated here and throughout this section are in relation to the
nearest segment of the present Rio Henares channel unless there is specific reference to
closer drainage lines.

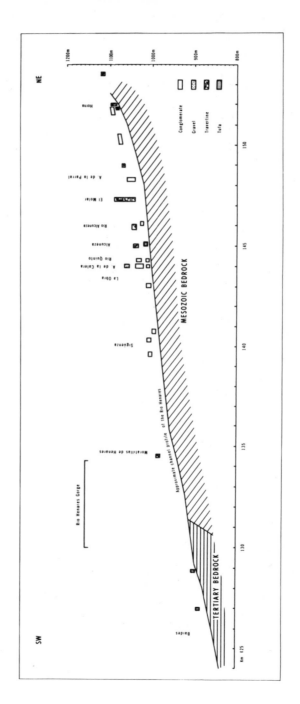

Fig. 32.--Longitudinal profile of the upper Rio Henares. The horizontal scale is given in relation to kilometer posts of the Madrid-Zaragoza railroad as stated on the Mapa Topográfico Nacional.

The 18 to 20 m Middle Terrace

Deposits related to the Middle Terrace are the most common of the upper Henares
terrace sequence. The typical facies include fluvial conglomerates and calcareous precipitates; these may be complexly interbedded, such as north of Horna (+25 m), or preserved
one without the other, which is the more usual case throughout the remainder of the upper
Henares. Middle Terrace conglomerates (+20 m) and travertines (+40 m) are preserved
above the Henares right bank between Horna and Mojares (see Fig. 32), and at Mojares
the El Molar middle travertine platform records this phase. Additional, prominent Middle Terrace deposits comprise the travertine platform at Alcuneza (+30 m) and define the
extensive conglomerate-gravel surface of the Arroyo de la Calera terrace sequence. Further documentation of the Middle Terrace abounds but the situations cited above are the
most informative.

Middle Terrace deposits record a protracted alluviation cycle. The widely scattered disposition of the terrace fragments indicates a broad, open, valley bottom of limited relief, laced by the river channel. Calcareous material precipitated in intrastream
situations, and perhaps in localized depressions or ponds; the elevation of these features
varied much in the same way that elevation varies along the modern valley floor. As fluvial deposits became indurated and cemented to form conglomerate, colluvial and detrital
material formed the breccia deposits that occur at elevations comparable to the Middle
Terrace, but not on top of its deposits.

The 6 to 8 m Low Terrace

Low Terrace deposits are not widely recorded or, with one exception, of morphologic prominence. At the Henares-Alboreca confluence, it is recorded by conglomerate
beds (+17 m) and two, broad, flat platforms (+15 m) cut across Keuper bedrock. Low Terrace conglomerates also occur at the Quinto-Henares confluence (+9 m), at La Obra
(+10 m), and above Sigüenza (+10 m). Loose gravel covers are not characteristic of this
terrace and occur only at the head of the Arroyo de la Calera. The paleosol developed
within this arroyo fill is partly coeval with the Low Terrace phase and, although direct
stratigraphic evidence is lacking, may be considered contemporary with tufa deposits at
Alcuneza (+10 m), and those at Horna (+15 m) which have yielded a Late Pleistocene radiocarbon date.

The Campiña Terrace

The flat valley bottom of the modern campiña, into which the Henares is incised
1 to 3 m, is the youngest fluvial surface of the upper Henares. Quartzite gravels mantle

this surface locally but most of the campiña is characterized by reworked, Ap soil horizons. Downstream from the Mesozoic-Tertiary bedrock contact, more aggressive incision of this surface (6-10 m), in response to an increase in the river gradient, exposes
deposits that underlie the surface and suggests that the upper Henares campiña surface
may obscure older, buried Low Terrace deposits. Whether or not the campiña surface
represents planation rather than the terminus to an aggradation phase in the upper Henares
valley is correspondingly unclear.

The Rio Salado System

The upper Salado Basin

The northwestern sector of the study area is drained by the Rio Salado system,
which joins the Rio Henares at Baides. The disturbed bedrock zone of the Mesozoic-
Tertiary geologic contact is cut through by the Rio Salado at Huérmeces. For more than
8 km upstream from the Huérmeces gorge, however, the courses of the Salado and its
major tributaries wind through constricted valleys that are in sharp contrast to the valley
bottom that characterizes most of the upper Salado drainage. These morphologic distinctions, rather than the geologic contrasts, serve as the basis here for distinguishing
between the upper and lower Salado. The elongated, saucer-shaped campiñas of the upper
Salado and its major left bank tributary, the Rio de la Hoz, are more extensive than those
of the upper Henares and, despite their proximity, the campiñas exhibit different elevation
ranges since they are graded to different temporary base levels set by their respective
gorges. The general elevation of the floor of the Henares campiña is 980 to 1080 m, while
that of the Rio de la Hoz is 920 to 980 m, and that of the Rio Salado is 930 to 1000 m. At
Olmeda de Jadraque, where the Rio Salado leaves the campiña, it is joined by the Rio
Cercadillo before it enters the gorge through the upland surface. Near El Atancé, the
Salado is joined by the Rio de la Hoz which cuts through this upland at Cirueches.

The general characteristics and relative occurrences of Pleistocene deposits in the
upper Salado are quite similar to those of the upper Henares: massive travertines and conglomerates, breccias, tufas, and gravels, ultimately derived from the Mesozoic bedrock.
In the preceding section, a sequence of three terraces above the campiña surface was
established for the upper Henares. The similarity of past and present sedimentary processes allows for discussion of the Salado sequence within the framework of the upper
Henares terrace sequence. In several cases, however, the record is significantly more
complete and these instances require more detailed discussion.

The Campiña Terrace

The extent of the campiña surface is approximately defined by the limits of the flat-bottomed valley floor. At its margins, the campiña gives way to gently concave changes of slope, offset below older and higher surfaces. The present channels of the upper Salado and its tributaries represent little more than incisions into the campiña, generally in the order of 1 m in the case of the Rio de la Hoz, and as much as 2 m in the upper Salado above Imón. Consequently, large portions of the campiña surface are still subject to seasonal inundation--in which cases the functional sectors are locally referred to as lagunas.

The campiña surface is distinctive morphologically but less so in terms of deposits. Hydromorphic soils are characteristic of the waterlogged laguna situations while organic paternia or borovina soils predominate elsewhere. Locally, as in the upper Rio de la Hoz, a thin cover of limestone gravel may mantle portions of this surface. These veneers prompt the consideration of the campiña surface as a terrace, even though in sensu stricto such a reference would not be appropriate since segments of the surface still experience inundation, and exposed sections of fluvial deposits have not been recorded.

The Low Terrace

The Low Terrace is gently offset 5 to 11 m above the general level of the campiña surface. Remnants of this terrace, occasionally capped with alluvial deposits, are commonly seen along the valley floor rising as much as 15 m above the campiña. The scattered distribution as well as the morphologic variance of these remnants indicate a once extensive fluvial surface later subjected to both downcutting and planation.[8] Alluvial deposits are not universal but nonetheless they demonstrate a fluvial origin. The Low Terrace exposures can be enumerated and briefly described as follows (see Fig. 33 for locations):

1) East of Salinas de la Olmedilla (930 m). Approximately 25 cm of unstratified, caliche-coated, subrounded gravel (Table 15) under a 40 cm, humic Ap-horizon, 8 to 10 m above the general elevation of the campiña surface.

2) East of Imón, Km 15.8 (937 m). Veneer of quartzite gravel 5 to 7 m above the campiña surface.

3) Imón (940 m). Massive tufa deposits 5 to 7 m above the Rio Salado.

---

[8]Platforms or benches are also prominent where campiñas and midslopes are cut into Keuper rocks. These features are marked by clear morphologic definition and altimetric grouping. They are preserved at elevations similar to fluvial platforms in their immediate proximity but they are without deposits. These plátforms almost certainly are of fluvial rather than structural origin but this remains inferential without sedimentological evidence.

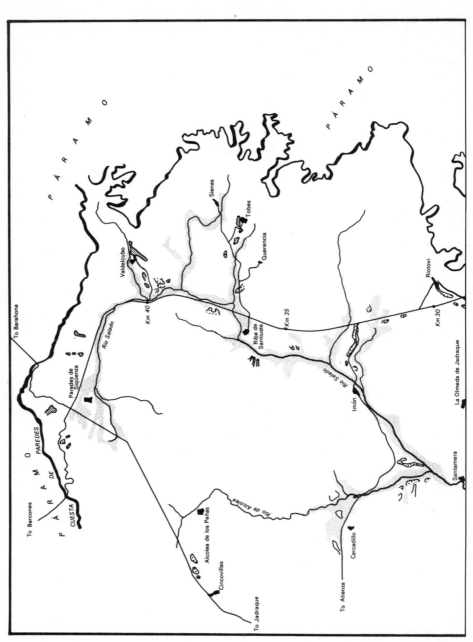

Fig. 33.--Pleistocene deposits

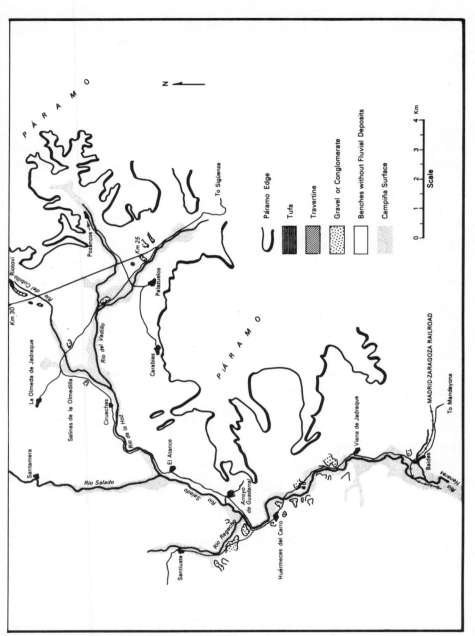

of the upper and lower Rio Salado

4) Northeast of Imón, left bank (942 m). Fractured but rounded quartzite gravels (Table 15), derived from Buntsandstein, in a partly indurated, sandy matrix 8 to 10 m above the Rio Salado.

5) Valdelcubo, left bank (975 m) and Querencia (965 m). Cemented conglomerate 10 to 11 m above the Rio Salado.

The lithology of the Low Terrace gravels and conglomerates depends upon the local situation. Quartz and quartzite gravels in the upper Salado are derived from massive Buntsandstein conglomerates that are only exposed in the Riba de Santiuste anticline. As a result, Pleistocene conglomerates near Imón and Riba de Santiuste have significant quartzite components. On the other hand, gravels in the Rio de la Hoz campina are composed almost exclusively of dolomitic or quartzitic carniolas limestones.

The Middle Terrace

Middle Terrace deposits are widely recorded throughout the upper Salado in the form of conglomerates and travertines. As in the case of the upper Henares, these conglomerates are situated above those of the Low Terrace so as to leave no doubt that each represents a separate and distinct cycle of alluviation. This conclusion is strengthened by the contrast of the conglomerate matrices, particularly in terms of color and cementation. The reddish-brown matrix of the Middle Terrace conglomerate indicates derived red soil sediment, while the Low Terrace conglomerate matrix is yellowish-red and less well cemented. These distinctions may be locally obliterated by facies changes, but the contrasts are sufficiently predictable to corroborate the morphologic and altimetric distinctiveness of the conglomerate terraces. In the upper Salado the vertical distinction of these terraces is amply demonstrated above Imón, and again between Riba de Santiuste and Valdelcubo. At Valdelcubo, for example, pockets of Low Terrace conglomerate and an extensive Middle Terrace travertine cap are offset 8 to 10 m one from the other (Fig. 34).

Middle Terrace conglomerates occur at the following locations (see Fig. 33):

1) East of Palazuelos (965 m). Conglomerate platform with occasional tufa-travertine litter, 15 to 17 m above the Rio de la Hoz.

2) Near Riotovi, between 985 and 1017 m. Conglomerate cap on remnants of an extensive platform 17 to 20 m above the Rio del Cubillo.

3) Km 31.5 (957) and Km 31 (960 m) of the Paredes-Alcolea road. Conglomerate blocks, perched on platform remnants with tufa-travertine litter, on the west side of the road, 27 to 30 m above the Rio Salado.

4) East of Imón (947 and 957 m). Conglomerate caps a platform 23 to 25 m above the Rio Salado right bank.

5) East of Cincovillas (1018 m). Conglomerate blocks cap a small knoll approximately 20 m above the headwaters of the Rio Cercadillo.

Other segments of the Middle Terrace are preserved by thick caps of laminated, calcium carbonate deposits along the left bank of the upper Salado, below the carniolas

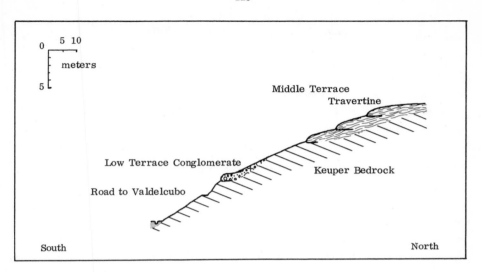

Fig. 34.--Transect at Valdelcubo. The Middle Terrace travertines comprise part of the expansive terrace platform at the village of Valdelcubo, while the Low Terrace conglomerates can be related to similar deposits farther south along the Rio Salado right bank (see Fig. 33).

limestone páramo escarpment. At Tobes, some 3 m of very well-cemented, pure calcite tufa rest on limonitic marl and are inclined 3° toward the Salado between 1030 and 1038 m. Similar accumulations are found farther upstream at Valdelcubo, between 995 and 1020 m. A third sequence of travertine deposits, discussed in greater detail below, is found east of Paredes de Sigüenza; the lower platforms of the sequence belong to the Middle Terrace.

The concurrent development of the Middle Terrace conglomerates and travertines is demonstrated by several factors. The basal facies of the lowest travertine deposits at Paredes and Valdelcubo are gravelly; the altimetric relationships of these facies coincide with an identical floodplain level as that of the conglomerates. Travertine accumulation appears to have been a local, intrastream phenomenon where carbonate precipitates were laterally or vertically interstratified with clastic fluvial beds. In such situations gravel and travertine deposits may be more or less contemporary, probably indicating protracted periods of polygenetic accumulation. Such development is suggested west of Valdelcubo where a small Middle Terrace travertine platform at 997-1001 m is littered with conglomerate debris including small (25 cm) conglomerate blocks; both the platform and the conglomerate material are quite distinct from the Low Terrace conglomerate (975 m).

Basal gravels appear to be absent from travertine deposits at Tobes; instead, marls represent this facies. The Tobes precipitates are peripheral to the ancient Middle Terrace floodplain and are somewhat higher than it. However, the inclination of the deposits demonstrates that they were graded to it.

The paleo-environment indicated by the disposition of these deposits is a broad, open, valley bottom, partly mantled with fanglomerates introduced from the Buntsandstein exposed in the Riba de Santiuste anticline as well as from upland limestones. The margins of this ancient alluvial surface were gently graded to the river channel. Throughout the extended period of time represented, lateral-entrant streams with water charged with carbonates from the limestone upland precipitated travertines locally, and along the immediate margins of the ancient Rio Salado such deposits graded into and interfingered with the gravels of the bed load. Detrital materials littering higher slopes were transported by colluvial agents and intermixed with fine surficial sediments to become cemented into breccias.

The contemporaneous formation of these breccias and of the Middle Terrace conglomerates is demonstrated by their relationship in the Rio de la Hoz campiña. Near Km 24 of the Paredes-Alcolea road, crudely stratified, subangular-to-subrounded, limestone detritus in a well-cemented, light reddish-brown (5YR 6/4) matrix rests on an elongated platform inclined 3° toward the valley bottom. The breccia deposits occur between 960 and 1025 m. This platform is a morphologic continuation of a Middle Terrace remnant (+15 m) recorded 1 km downstream on the right bank of the Rio de la Hoz. The composite profile of these two features defines a continuous surface graded toward an ancient floodplain. The morphometry of the clastic capping deposits changes downslope, suggesting a transition from angular waste to rounded pebbles. Altimetric relationships between these breccias and the Middle Terrace remnants elsewhere in the upper Salado also point to lateral interdigitation.

Additional breccia deposits are found east of the Paredes-Alcolea road at Km 32 (959 m) only several meters higher than adjacent Middle Terrace conglomerates, and north of the Tobes travertines, perched on a Keuper knoll at 1032 m. Massive breccias also occur above the Middle Terrace travertine platform (ca. 1035 m) at Paredes.

The High Terrace

Throughout the upper Salado drainage residual forms, mainly without deposits but higher than the Middle Terrace platform, attest to an older floodplain surface within the valley confines. Sufficient sedimentological evidence exists to demonstrate the development of a high terrace.

At Riba de Santiuste, the Salado has cut through the Muschelkalk hogbacks along the flanks of the exhumed anticline. A right bank 80 m platform 1 km downstream is inclined 4° toward the river. The surface (1020 m) cuts dipping (10°) Buntsandstein strata and is veneered with quartzite gravel. These deposits are higher than other Pleistocene features in the valley downstream and east of Imón, and more than 35 m above the Middle

Terrace 2 km farther upstream near Riba de Santiuste. This fluvial platform was obviously graded to a higher floodplain and is, therefore, a vestige of a High Terrace.

Travertines above the valley floor at Paredes de Sigüenza also relate to this older period of alluviation. Two travertine platform levels can be identified east of the village; the lower one already has been cited as relating to the Middle Terrace. The upper platform level at 1058 to 1070 m is a High Terrace remnant, as is the travertine platform at approximately 1080 m above Km 39 of the Cuesta de Paredes to Barcones road (see Fig. 33).

Unlike the upper Henares, the upper Salado provides several sites where long and detailed sedimentary sequences are recorded. Two of these exposures occur within the morphological entity of the Middle Terrace--at Riba de Santiuste and at Riotovi--while a third sequence, recorded near Paredes, demonstrates the stratigraphic relationship of Middle and High Terrace deposits. These exposures illustrate the details of Pleistocene geomorphological evolution and warrant a more complete discussion.

Riba de Santiuste

Conglomerates associated with the Middle Terrace rest on two benches above the Rio Salado north of Riba de Santiuste (p. 133 and Fig. 33). These truncated spurs, immediately west of the Alcolea-Paredes road at Km 37.5, have been partially quarried to recover gravels (Fig. 35). Exposures on the east and north faces of one of these spurs reveal different aspects of the series of sediments. The east face exposure is referred to here as the Horizontal Sequence, while the steeply inclined beds of the north face are referred to as the Inclined Sequence.

The Horizontal Sequence can be outlined as follows, from base to top (see Fig. 36 and Fig. 37):

> a) Over 340 cm of subrounded, medium coarse gravels (Table 15) in a reddish-yellow, coarse sand matrix (Table 7, No. S120). The well-stratified gravel is comprised of quartzite, dolomite, and shale rocks derived from the Paleozoic and Mesozoic material exposed within the Riba de Santiuste anticline; finer gravels in the topmost deposits suggest limited sorting. The uppermost 5-15 cm of the beds are cemented by a laminated, recrystallized, yellow (10YR 7/6), medium coarse-to-coarse, sandy, calcrete crust. The gravel body of the deposits is indurated in the medium coarse sand matrix and locally interrupted by sand lenses. A 35-50 cm sand lens can be identified at the base of the deposits but the overall thickness is obscured by the floor of the excavation pit. An upper 20-25 cm, light reddish-brown (5YR 6/4), medium-sandy clay lens contains shell fragments (rare) and is characterized by current-bedding. Clear, straight contact; disconformity.
>
> b) 70-120 cm of light brown, sandy marl (Table 7, No. S122). The laminated sediment is weakly consolidated, with coarse prismatic structure and bands of limonitic oxidation stains. Small shell fragments also occur. Diffuse, irregular, involuted contact.

Fig. 35.--Middle Terrace at Riba de Santiuste (view looking southwest). The Inclined Sequence is in the center of the photograph and the Horizontal Sequence to the left of center.

c) 70-150 cm of light gray marl (Table 7, No. S123). The body of the medium coarse, angular blocky to prismatic marl is weakly consolidated, with oxidation staining. Unidentified shell and shell fragments abound in the uppermost sediment. Clear, smooth contact.

d) 35-90 cm of reddish-yellow, sandy marl (Table 7, No. S124). The weakly consolidated, stratified sediment shows wavy or undulating bedding with thin bands of limonitic oxidation. Occasional, shell fragments and traces of organic material occur throughout. Fine gravel lenses may be found locally, in addition to a 1-2 cm, light reddish-brown (5YR 6/4), silty clay lens. The laminated clay lens is medium angular blocky-to-columnar in structure and contains occasional shell fragments. Abrupt, irregular contact; disconformity.

TABLE 7

SEDIMENT CHARACTERISTICS OF THE UPPER SALADO MIDDLE TERRACE
AT RIBA DE SANTIUSTE, RIOTOVI, AND PAREDES DE SIGÜENZA

Sample Number	Strata	Color	Texture	Sorting	CaCO$_3$ % wgt	pH
			Riba de Santiuste Horizontal Sequence			
S120	Bed a	7.5YR 6/6	Coarse sand	Moderate	73.7	8.4
S122	Bed b	7.5YR 6/4	Sandy marl	Moderate	50.2	7.7
S123	Bed c	5Y 7/2	Marl	Moderate	43.1	7.7
S124	Bed d	7.5YR 6/6	Sandy marl	Moderate	47.3	7.7
S126	Bed e	7.5YR 6/6	Coarse sand	Poor	76.6	7.8
			Riba de Santiuste Inclined Sequence			
S99	Bed a	10YR 7/2	Marl	Poor	44.4	7.7
S100A	Bed b	10YR 6/8	Coarse sand	Moderate	41.3	8.0
S98	Bed c	7.5YR 6/6	Coarse sand	Moderate	67.7	8.0
S97	Bed d	10YR 7/4	Sandy marl	Poor	54.4	7.6
S104	Bed e	6YR 6/6	Coarse sand	Poor	3.6	7.6
S105	Bed f	6YR 6/6	Silty fine sand	Poor	51.1	7.5
S107	Bed g	5YR 6/6	Coarse sand	Poor	6.3	7.7
S109	Bed h	7.5YR 6.5/6	Sandy marl	Poor	12.8	7.7
S110	Bed i	5YR 6/4	Sandy marl	Moderate	57.8	7.5
S111	Bed j	10YR 7.5/1	Sandy marl	Moderate	54.2	7.3
S96	Bed k	10YR 7/6	Sandy marl	Moderate	54.6	7.8
S113	Bed l	5YR 6/6	Coarse sand	Poor	7.9	7.5
S115	Bed m	5YR 6/6	Silty sand	Moderate	57.4	7.5
			Riotovi			
S204	Bed a	2.5Y 7/4	Coarse sand	Good	70.5	7.5
S215	Bed b	2.5Y 7/2	Sandy marl	Moderate	30.4	7.2
S203	Bed c	5Y 7.5/4	Sandy marl	Moderate	71.1	7.8
S213	Bed d	10YR 8/3	Sandy marl	Moderate	56.8	7.6
S212	Bed e	10YR 7/3	Marl	Moderate	68.7	7.4
S216	Bed f	2.5Y 7/2	Marl	Moderate	59.6	7.6
S208	Bed g	2.5Y 7/4	Coarse sand	Poor	26.1	7.7
			Paredes de Sigüenza			
S311	Bed a	7.5-10YR 6/3-4	Sandy marl	Poor	32.1	7.6
S310	Bed b	5YR 5/6	Coarse sand	Poor	30.4	7.5

e) Over 200 cm of subrounded, medium coarse gravels (Table 15) in a reddish-yellow, coarse sand matrix (Table 7, No. S126). The gravel body is crudely stratified and unsorted, but localized lenses of current-bedded gravels and pockets of pyrolusite-stained gravel are evident. The topmost 100-150 cm of the gravel is cemented by calcareous material to form genuine conglomerate.

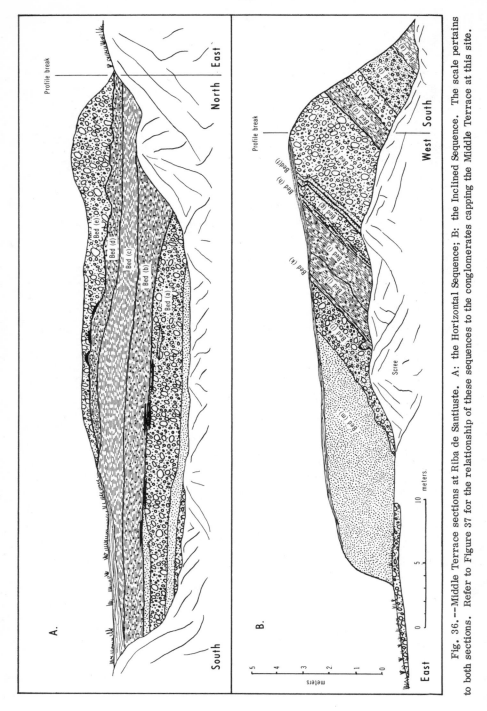

Fig. 36.--Middle Terrace sections at Riba de Santiuste. A: the Horizontal Sequence; B: the Inclined Sequence. The scale pertains to both sections. Refer to Figure 37 for the relationship of these sequences to the conglomerates capping the Middle Terrace at this site.

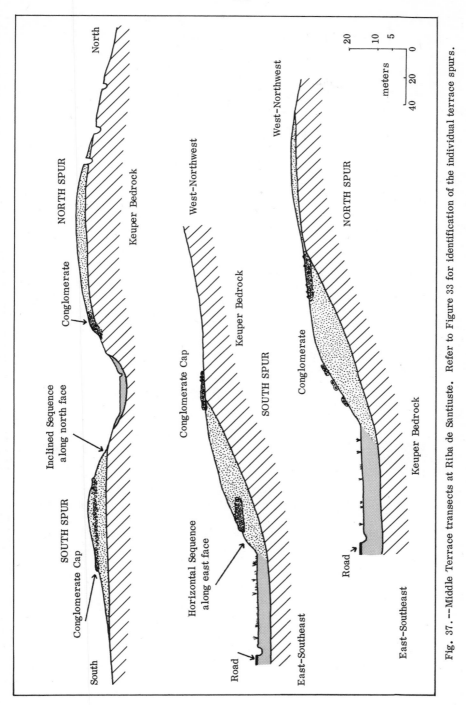

Fig. 37. -- Middle Terrace transects at Riba de Santiuste. Refer to Figure 33 for identification of the individual terrace spurs.

The upper cemented gravel of the Riba de Santiuste Horizontal Sequence can be followed west from the east face exposure to the floor of the north face pit, where it forms a conglomerate pavement. This pavement can be traced to within 4 m of steeply dipping conglomerate beds of the Inclined Sequence, before it becomes buried under the gravel pit floor. Conglomerate strata comprise an upper facies strata of the Inclined Sequence also, but the overall thickness of this sequence is considerably greater. The north face Inclined Sequence may be summarized as follows (from base to top):

a) Over 120 cm of light gray marl (Table 7, No. S99). The basal contact of this weakly consolidated sediment cannot be located, but it is presumed to rest on the Keuper shale. The marl has weakly platy structure and contains shell fragments and oxidation stains. Clear, wavy contact.

b) 210 cm of fine, subrounded pebbles in a brownish-yellow, coarse sand matrix (Table 7, No. S100A). The pebble materials are dispersed throughout the sandy matrix, which has undergone partial oxidation and contains small shell grit. The upper level of the sand is interlayered with a 20 cm, weakly platy, white (5Y 8/2), sandy marl, with pockets of oxidation locally. Abrupt, wavy contact.

c) 90 cm of crudely stratified, subangular to subrounded gravel in a reddish-yellow, coarse sand (Table 7, No. S98). Rare pieces of calcareous plant stem casts may be found in the loose, coarse sand matrix. Abrupt, wavy contact.

d) 150 cm of clayey, very pale brown, sandy marl (Table 7, No. S97). The sand is weakly consolidated and has medium-weak, granular structure. Limonitic oxidation stains occur throughout the sediment. Abrupt, wavy contact; disconformity.

e) 330 cm of well-stratified gravel in a reddish-yellow, coarse sand matrix (Table 7, No. S104). The subrounded, medium gravels (Table 15) occur throughout the loose, sandy matrix, which is partially cemented in the basal 2-4 cm and is interlayered with 7 cm sand lenses in the uppermost transitional sediments. Clear, wavy, convoluted contact.

f) 70 cm of reddish-yellow, fine sand (Table 7, No. S105). The lowermost 15-25 cm of this sediment consists of a calcareous, fine sand body which grades vertically into 25-35 cm very pale brown (10YR 7/3), sandy, marl-like sediment (46% carbonates). The uppermost 20 cm consists of sand similar to that of the basal portion of the deposits. Abrupt, wavy, convoluted contact; disconformity.

g) 70 cm of well-stratified gravels in a reddish-yellow, coarse sand matrix (Table 7, No. S107). The subrounded gravels (Table 15) are similar in lithology and morphometry to those in unit (e) below. Clear, wavy contact.

h) 50 cm of reddish-yellow, coarse sand (Table 7, No. S109). The body of this sediment is interspersed with several small (1-2 cm), gravel lenses. The uppermost 5-7 cm are cemented by a reddish-brown (5YR 6/3.5), medium coarse sandy calcrete crust containing occasional subrounded pebbles. Clear, wavy contact; disconformity (?).

i) 170 cm of reddish-brown, sandy marl (Table 7, No. S110). The body of the sediment is essentially uniform but with some color change and oxidation staining. Current-bedded medium-to-fine sand occurs locally. Diffuse, irregular contact.

j) 130 cm of light gray, sandy marl (Table 7, No. S111). The marl has medium coarse angular blocky-to-prismatic structure and contains shell fragments and red (2.5YR 5/6) oxidation nodes. Very pale brown (10YR 7/4), clayey, fine sand bands occur in the upper portion of the deposits. Clear, smooth, convoluted contact.

k) 30 cm of yellow, sandy marl (Table 7, No. S96). The weakly consolidated sediment has fine platy structure and lighter, pale yellow (5Y 7/3) bands of limonitic oxidation. Shell fragments occur locally within the stratified, current-bedding. Abrupt, wavy, convoluted contact; disconformity.

l) 170-200 cm of subrounded, medium-grade gravels (Table 15) in a reddish-yellow, coarse sand matrix (Table 7, No. S113). The gravel body is well stratified, locally current-bedded, and with pockets of pyrolusite stained gravels. The upper 5-20 cm of the gravel becomes cemented into a calcareous crust. Abrupt, wavy, convoluted contact.

m) Over 360 cm of reddish-yellow, silty sand (Table 7, No. S115). The sediment is well stratified, semiconsolidated and fairly homogeneous but contains a larger fine sand fraction and lightens to a pink (6YR 7/4) toward the uppermost layers. Bedding varies systematically; basal sediments, resting upon and partially interworked with the underlying gravel, share the same steep dip in the order of 25°. The concave aspect of the bedding decreases vertically upward so that the beds are inclined only 10°-12° in their middle zone and as little as 2° in the topmost layers.

The entire Inclined Sequence dips east 28°-30° but the bedding direction of individual members differs and suggests a relationship between the overall texture of individual beds and bedding direction. Gravels were transported from the Paleozoic and Mesozoic conglomerates comprising the anticline core (Fig. 38), while bedding direction of the fine deposits (sands, silts, marls), although more varied, points to alluviation associated with the stream activity of a broad floodplain. Elaborate inferences on the basis of bedding direction are not justified, however, since at least in the case of the Inclined Sequence, the bedding is not primary. This is demonstrated by several factors. In the first case, the nearly uniform thickness of individual beds, the similar angle of inclination of all beds en échelon, and the nearly constant steep angle of dip (Fig. 39) are difficult, if not impossible, to explain if the beds are assumed to be undisturbed. Explanation of the inclined stratification of the finer materials, specifically fine sands and marls, is particularly elusive by that assumption. Second, alluvial origins are indicated by the several examples of current-bedding and the composition of the body of the deposits, so that colluvial grèzes litées are not indicated, even though cold climate features are recorded within the sequence. A third indication of disturbed bedding is provided by the relationship of the Inclined and Horizontal sequences.

As noted previously, the uppermost gravel-conglomerate of the Horizontal Sequence (Bed e) can be followed to the vicinity of the uppermost gravels (Bed l) of the Inclined Sequence (see Fig. 36). In addition, these gravels have distinctive characteristics, such as pyrolusite staining and current-bedding, sufficiently unique to distinguish them from all of the other gravel beds of both sequences. It is evident, therefore, that these deposits are lateral variants of the same facies. In addition, the sedimentology of the underlying series of each sequence is markedly similar, with Beds (d), (c), (b) and (a) of the Horizontal Sequence corresponding to Beds (k), (j), (i), and (g) and (h) of the Inclined Sequence respectively. The lateral conformability of these exposures demonstrates,

Fig. 38.--Riba de Santiuste basal gravels. The subrounded gravels and current-bedded 20-25 cm sand lense comprise Bed (a) of the Horizontal Sequence.

Fig. 39.--Riba de Santiuste convolution. These deposits include Bed (e) (lower right corner) through Bed (i) (upper left corner) of the Inclined Sequence.

therefore, that the bedding of the Inclined Sequence is not primary. The several meters of silty sand overlying both sequences was deposited while these beds were disturbed, as shown by vertically decreasing inclination of the stratification.

The sediments of the Riba de Santiuste exposures are embanked against Keuper spurs that rest upon the Muschelkalk hogback. The Pleistocene materials comprise a terrace 25 m above the Rio Salado. As much as 2 m of fine-to-medium coarse, crudely stratified or stratified quartzite and limestone conglomerate material caps this spur (980 m) and the adjoining spur to the north (Fig. 37). These cemented deposits in part overlie the inclined sediments and in part Keuper bedrock and indicate a disconformity of uncertain age and duration between the Middle Terrace conglomerate and the Riba de Santiuste sequence. The sediments derived essentially from within the exhumed anticline west of the site and alluviated by streams that drained this anticline and sluiced their load out onto the ancient Rio Salado floodplain. The river drainage at that time may have been to the north, forming headwaters of the Rio Alboreca.

The great total thickness of these deposits shows that the ancient Middle Terrace floodplain--30 meters or more higher than the present campiña--was a broad, aggrading valley bottom affording a variety of local sedimentation situations. The areal differences confound the establishment of stratigraphic equivalents whereby colluvial silts, marls, gravels or even conglomerates may be contemporaneous valley bottom deposits. The paleo-setting is one, therefore, of an open, expansive, flat-bottomed campiña with local slough and marsh conditions, marginal gravel fans, and continually or frequently changing drainage channels. Throughout the course of a protracted climatic interval, a given location may have experienced several distinctive sedimentological settings without intervening disconformities. This interpretation implies that facies changes need not have climatic implications; a sand lense within a gravel facies, for example, may record a drainage channel shift and mid-channel bar or riffle deposition quite apart from any changes in discharge regimen with climatic implications.

The morphogenetic implications of the Riba de Santiuste Middle Terrace deposits, based upon the Inclined Sequence, may be outlined as follows (from bottom to top):

a) Accumulation of marl precipitates, perhaps in shallow, impounded floodplain ponds, with seasonal inundation (Bed a).

b) Extended alluviation of pebbly channel sands, with periodic dessication (marls) (Bed b).

c) Alluviation of gravels by streams with increased competence (Bed c).

d) Deposition of sandy alluvium (Bed d).

e) Erosional disconformity.

f) Protracted accumulation of stream gravels under more moist conditions (Bed e).

g) Sedimentation of flood sands and marl (Bed f).

h) Erosional disconformity?

i) Alluviation of channel gravels with increased stream competence. Colder conditions with involuting of underlying beds (Inclined Bed g, Horizontal Bed a).

j) Floodplain sedimentation of sands and fine gravels, becoming drier (crusts) and perhaps colder (Inclined Bed h, Horizontal Bed a).

k) Erosional disconformity.

l) Alluviation of fine alluvium (Inclined Bed i, Horizontal Bed b).

m) Uninterrupted alluviation of marl precipitates in pond-like situations; cold (Inclined Bed j, Horizontal Bed c).

n) Fine floodplain sedimentation (Inclined Bed k, Horizontal Bed d).

o) Erosional disconformity.

p) Deposition of fluvial gravels; initially cold with soil-frost phenomena; becoming drier, with calcareous cementation (Inclined Bed l, Horizontal Bed e).

q) Protracted sedimentation of fluvial sands, initially under cold conditions recorded by soil-frost phenomena and wholesale dislocation of the frozen deposits of the Inclined Sequence (Bed m).

r) Erosional disconformity.

s) Floodplain alluviation of stream gravels, followed by calcareous cementation (Capping conglomerate).

Riotovi

At the village of Riotovi, conglomerate deposits are found atop remnants of a dismantled Middle Terrace platform (985-1017 m) on both sides of the Paredes-Alcolea road at Km 29.2. The south face of the uppermost spur is transected by a road to the village so that several sedimentation sequences are exposed. The complete sequence (see Fig. 40) can be outlined as follows (base to top):

a) 450 cm of crudely stratified, subrounded gravel (Table 15) in a pale yellow, coarse sand (Table 7, No. S204). The gravel rests on Keuper marl and includes some limestone and quartzitic or dolomitic limestone cobbles and boulders up to 30 cm in diameter. The consolidated sand matrix shows local limonitic staining and contains a 25 cm non-homogeneous, platy, gray (10YR 6/1), sandy-to-stoney marl, also with oxidation stains. Clear, smooth contact; disconformity.

b) 30 cm of light gray, sandy marl (Table 7, No. S215). The strongly structured, medium coarse, blocky sediment has gritty, calcareous laminae and limonitic stains. Clear, wavy contact.

c) 90 cm of pale yellow, sandy marl (Table 7, No. S203). The massive non-homogeneous deposit is characterized by thick (4 cm), limonitic horizons and light reddish-brown (2.5YR 7/4), calcareous, coarse sand and gravel in lenses up to 10 cm thick. Clear, irregular contact.

d) 50 cm of very pale brown, sandy marl (Table 7, No. S213). This weakly cemented sediment occurs as an undulating lens with some thicker pockets and is partially interbedded with underlying Bed c, the limonitic bands of which tend to follow the base of this bed. The irregular basal contact of Beds c and d suggests involutions. Obscured upper contact; disconformity (?).

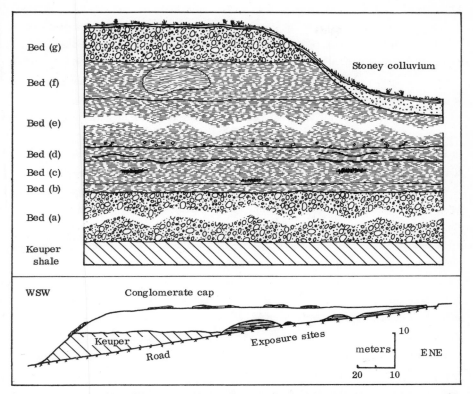

Fig. 40.--Composite section of Middle Terrace deposits at Riotovi. Aspects of the facies are exposed at the several sites indicated in the transect at the bottom.

e) 250 cm of very pale brown marl (Table 7, No. S212). This is a consolidated-to-semicemented sediment with shell fragments and occasional basal, subrounded, medium-grade gravels. Gradual, wavy contact.

f) 150 cm of pale yellow marl (Table 7, No. S216). The body of this sediment is massive and amorphous, but in pockets becomes strongly indurated to form a very pale brown (10YR 8/3), dense marl with shell fragments. The marl facies locally grades vertically into 35 cm of well-cemented, non-porous, white (10YR 8/2) tufa, also with shell fragments and containing occasional, subrounded pebbles. Clear, irregular contact.

g) 200 cm of crudely stratified, subrounded gravel (Table 15) in a pale yellow, coarse sand (Table 7, No. S208). The gravel is comprised of limestone, quartzitic or dolomitic limestones, and derived travertine and laterally is interdigited with a thick (45 cm) layer of light gray (2.5Y 7/2), sandy marl containing abundant shell fragments. The irregular basal contact suggests convolutions. Both the gravel and the marl show strong oxidation staining in their upper parts which locally become consolidated by a tufaceous crust.

The Riotovi sequence is a composite of several exposures along the road to the village. The lowermost exposure consists of Beds (a) through (d), which rest on Keuper

bedrock and dip 1° to 2° to the southeast. Beds (e) through (g) are exposed 47 m farther
up the road where they dip 2° to the west-southwest, as do the conglomeratic beds that
cap the platform. Outcrops of certain facies of the overall sequence can be observed
along most of the length of the road but intervening segments are buried by slope debris or
manure and cannot be traced directly (see Figs. 41 and 42). The slope debris comprises
2 to 3 m of unconsolidated, very pale brown (10YR 7/4), calcareous, partly stratified but
variably bedded, stoney rubble. Lithologic composition is the same as the conglomerate
cap of the spur and identical with material on the north flank of the spur that has weathered from the conglomerate. This colluvial debris is considerably younger, therefore,
than the underlying beds described above. Overlying the colluvial rubble are 30 to 50 cm
of very pale brown-to-brown (10YR 7/4-5.5/3), calcareous, sandy silt. This slightly
humic material at places overlies Bed (g) with a clear unconformity while at the same
time the contact with the colluvial debris is less clear and more transitional. The sediment indicates a stripped soil mantle, perhaps a meridional braunerde now on calcareous
sediments, although exact determination may be impossible because of probable more
recent humification, including that from overlying or upslope manure piles.

Certain analogies between the Riba de Santiuste and Riotovi sequences are apparent. Both exposures occur within features capped by the Middle Terrace conglomerate
and record the interdigitation of gravel and marl beds. In addition, the cold-climate involutions noted in each case are distinctive. The Riotovi exposures record, therefore, a
facies variation of the Riba de Santiuste type-site. Most probably, the Riotovi deposits
are contemporaneous with Beds (a) through (l) at Riba de Santiuste, determined primarily
on account of the disconformities, involutions, and calcrete-type crusts recorded in each
case.

In summary, the Riotovi deposits record the following events, from bottom to top:

a) Erosion of the Keuper marl bedrock and alluviation of coarse gravels (Bed a),
with local precipitation of marls under conditions of partial oxidation.

b) Erosional disconformity.

c) Marl deposition and oxidation (Bed b).

d) Marl deposition and oxidation, with greater stream alluviation (Bed c).

e) Sedimentation of fine alluvium, probably under cold conditions (Bed d).

f) Protracted marl deposition, initially in association with stream alluviation
(Bed e).

g) Sedimentation of lacustrine-like marl, as well as some tufa with pebbles, indicating marginal stream situations (Bed f).

h) Extended alluviation of gravels and marl, with oxidation. Initial cold conditions
superseded ultimately by a high water table (Bed g).

i) Erosional break or hiatus preceding alluviation of the conglomeratic cap.
Direct contacts are not observable.

137

Fig. 41.--Middle Terrace deposits at Riotovi. Beds (a) through (d) are shown.
The weakly cemented sandy marl of Bed (d) establishes the topmost deposit at the right.

Fig. 42.--Middle Terrace deposits at Riotovi. An exposure of Beds (f) and (g) is
shown.

j) Partial erosion of the conglomeratic cap and accumulation of slope debris.

k) Soil development (?) of uncertain age prior to an erosional period.

Paredes de Sigüenza

A somewhat different aspect of alluviation, at least in part associated with the Middle Terrace, is recorded near Paredes de Sigüenza, north of Km 43.2 of the Paredes-Alcolea road. Here massive travertines (Fig. 43) have preserved three separate platforms at different elevations above the valley floor (Fig. 44). The lowermost platform (1030-1035 m) comprises 3.5 m of white (10YR 8/2), well-cemented, slightly porous tufa with some shell fragment. The middle platform (1041-1046 m) consists of 3.5 m of very well-cemented, light brownish-gray (10YR 6/2) travertine. This contains coarse sand particles and shell fragments and in the basal portions, angular-to-subrounded gravels and detrital material. The travertine overlies as much as 50 cm of sandy marl which vertically varies in color from very pale brown (10YR 7/3)-to-pink (6YR 7/4), to light brown (7YR 6/4). The body of the sediment contains some detrital material and small root casts and calcareous bands. The uppermost platform is capped by 4.5 m of very well-cemented, non-porous, light yellowish-brown (10YR 6/4), nearly pure calcite travertine. Below the west edge of this platform (1058-1070 m), over 60 cm of cemented, slope-breccia is preserved intact 3 m below the travertine-Keuper contact.

The road cut at the base of the lowest platform exposes several meters of inclined deposits on top of the Keuper bedrock (Fig. 44). The sediments are found between 1000 and 1003 m and may be summarized as follows (from base to top):

a) 120-140 cm of light-to-pale brown, sandy marl (Table 7, No. S311). The loosely consolidated, coarse angular blocky sediment is interlayered with a 10-12 cm lens of indurated, fine angular detrital material and some well-cemented, light brown (7.5YR 7/4), travertine-like crusts, dipping approximately 20°. Subrounded limestone gravels occur sparsely dispersed throughout the basal portion of the deposits. Abrupt, wavy contact.

b) 220 cm of subangular chips (Table 15) in a yellowish-red, coarse sand (Table 7, No. S310). The deposits dip southeast 38° to 40°, are well stratified, and comprise limestone, dolomite, quartzite and shale lithologies. Calcareous cementation varies from weak to well cemented.

Most probably several distinct periods of travertine accumulation are recorded by the platforms above Paredes de Sigüenza. The highest platform records the oldest phase. This was followed by an interval of erosion and partial destruction of the travertine prior to the second depositional phase that formed the middle and lower platforms in riverine or marginal floodplain situations, as shown by the underlying pebbly marl and detrital character of the basal travertine deposits. This terrace level subsequently was partially eroded and at some undetermined point in time cementation of slope materials followed to form the breccia deposits recorded below the high platform. The steeply inclined detrital

Fig. 43. --Terraces at Paredes de Sigüenza

beds at the base of the lowest platform also may relate to this same period. The slightly
gravelly, basal sandy silt is alluvial in origin but the overlying stratified and cemented
scree is characteristic of grèzes litées or éboulis ordonnés-type slope deposits formed
under periglacial or cold-climate conditions. The depositional setting was therefore one
marginal to a stream bed or floodplain that experienced both alluvial and colluvial deposition. The high travertine platform is held to relate to the High Terrace while the middle
and lower platforms are considered to be related to the widely recorded Middle Terrace
phase. The temporal relation of the slope deposits is uncertain.

140

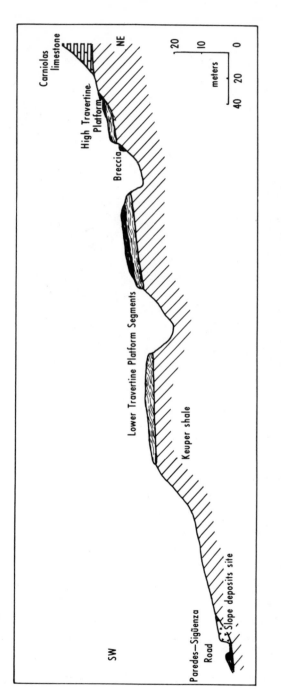

Fig. 44.--Transect at Paredes de Sigüenza. The two platform levels are distinguished by breccias found only in association with the high travertine platform and an underlying marl in the case of the two lower platforms. The éboulis ordonnés slope deposits described in the text are recorded at the left of the transect.

The lower Salado Basin

Pleistocene deposits of the lower Salado Basin are less informative in terms of stratigraphic details and paleoenvironmental implications. This is due to the fact that the sedimentological record is largely confined to surficial deposits and lacks deep, complex exposures. In addition, the definite recognition of Pleistocene terrace deposits is compromised in areas of Tertiary bedrock terrain. The calcareous Upper Oligocene quartzite, quartz, and limestone conglomerate prevalent downstream of Huérmeces bears a close resemblance to Pleistocene conglomerates of the Mesozoic landscape, so much so that unequivocal distinction is seldom possible. Consequently, these deposits cannot be used for correlation purposes unless clear-cut fluvial benches with diagnostic materials provide altimetric corroboration. As a result, interpretation of the widespread valley-side and midslope benches that are without deposits suggests a more complex sequence of alluviation cycles than can actually be demonstrated (see Fig. 33).

Upstream from Huérmeces, however, surface form is preserved in Mesozoic bedrock so that the conglomerates relate to Pleistocene alluviation. At the Cercadillo-Salado confluence, for example, a Low Terrace at 5 to 10 m is recorded by conglomerate above the right bank (920-925 m), and quartzite gravels mark the higher Middle Terrace 33 m above the left bank. Morphologic vestiges of these same levels are found near the village of Cercadillo at 6 to 10 m and 25 to 30 m respectively.

A suite of terraces upstream from the Regacho-Salado confluence best illustrates the succession of fluvial platforms along the lower Salado. Above the left bank of the river, a truncated valley entrant fan at the Fuente del Guadarral is sealed by massive tufa deposits (Fig. 45). The highest segment of the platform is 21 m above the campiña surface and retains a veneer of tufa fragments with some gravel. Immediately adjacent to this, a lower platform segment 14 m above the campiña surface, is composed of 3.5 m of massive tufa. Below these steps, the derived, whitish, tufa debris spews out upon the campiña in the form of a gently convex fan 5 to 7 m thick at its apex. The Fuente del Guadarral has apparently been active during several hemicycles of alluviation: 1) the tufa detritus litter of the 21 m surface records the oldest activity, 2) this was followed by partial dissection of this surface and subsequently, 3) by accumulation of massive spring deposits at 14 m, and 4) a final phase of erosion of the lower level and the deposition of the tufa fan graded to the campiña surface.

On the right bank of the modern Salado-Regacho confluence, gravel deposits demonstrate at least two distinct fluvial platforms. The lower surface (925 m) is cut across Keuper shale 30 m above the campiña surface. More than 1 m of subrounded, quartzite gravels with some boulders, in a sand matrix of reddish-brown color mantle this platform. Higher gravel-veneered surfaces--directly above the lower platforms--are cut into the

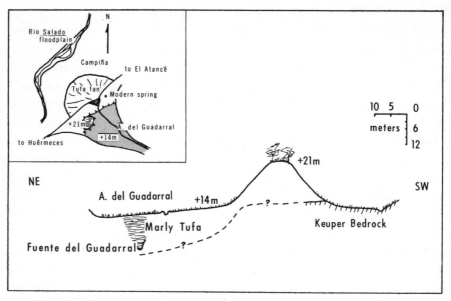

Fig. 45.--Transect at the Arroyo del Guadarral. Elevations of the fluvial surfaces are 14 m and 21 m above the campiña level.

same truncated interfluves at 955 m and define an older terrace some 61 m above the campiña. It may be noted that the modern floodplain of the Rio Regacho in this area is choked
with similar bed gravels derived from the Buntsandstein and Paleozoic bedrock.

Below Huérmeces the campiña surface widens out onto a flat-floored, cultivated
valley into which the Rio Salado is incised 2 to 4 m. Meandering has locally built a sandy
or gravelly floodplain as much as 80 m wide, large portions of which are aperiodically
active. A 1.5 to 2 m platform lies within this floodplain, experiencing less frequent inundation and little or no remodeling. Valley entrant fans along this lower segment of the
river are graded to the campiña surface, and the drainage channels that dissect these features grade to the present river channel. Local exposures in the campiña surface indicate
variable facies that include gray marls, brown sandy silts, and buried humified horizons.
At Baides, the 4 to 7 m campiña surface of the lower Salado merges with the Low Terrace
of the Rio Henares.

Altimetric and morphologic expression of a Low Terrace, at 15 to 20 m above the
river bed, contrasts with counterparts in the upper Salado and Henares Basins, which
have an elevation of 10 m and rarely retain surface form definition.[9]

[9] A singular exception at the Alboreca-Henares confluence (15 m) reflects a higher,
local base level resulting from the constriction of the valley at Alcuneza.

Approximately 2 km upstream from Viana de Jadraque, the road to Huérmeces traverses Low Terrace deposits 15 m above the Salado river bed. The exposure comprises about 1.3 m of crudely stratified, medium-to-coarse, rounded quartzite and limestone gravels (Table 15) in a silty sand matrix. A thin, calcareous cement encrusts the upper portion of the gravel. Vestiges of the Low Terrace reappear approximately 1 km farther downstream where they cut across steeply inclined Oligocene beds; gravel litters a left bank surface at 16 m. Alluviation of these spurs was probably contemporaneous with the tufa deposits at 14 m already described from the Arroyo del Guadarral.

Neither the Middle nor High Terraces are recorded by fluvial deposits south of Huérmeces, although a number of platforms presumably relate to the terrace stages recorded near the Salado-Regacho confluence. Middle Terrace benches are cut across inclined bedrock along the east bank 28 to 30 m above the river at 880 m and 886 m, 1 km and 2 km respectively upstream from Viana de Jadraque, which itself may be on such a platform (+ 28 m). Northeast of Baides a 27 m platform records this level near the Salado-Henares confluence at 870 m. Benches related to the High Terrace can be noted west of the Salado, both south of Huérmeces at 910 m and west of Viana de Jadraque at 920 m, 50 m and 70 m above the river respectively.

Summary

A suite of three terraces can be recognized within the Salado drainage system. As in the case of the upper Henares, the generalized altimetric range for each is a consequence of the varied remoteness of terrace remnants from present river courses, and of the specific portion of the original terrace that is preserved--ranging from the lower, channel deposits of the ancient stream to the highest, most peripheral valley margin colluvia. [10] The younger the terrace, the less the elevation disparity, due in part to the decreasing width of the younger floodplains, and in part to the fact that the period of aggradation and lateral planation appears to have been less complex and possibly more brief in the cases of the Low Terrace and the Campiña Terrace. Local grade and base level conditions at the time of alluviation also are reflected in the elevation differences.

The 40 m High Terrace

A High Terrace is recorded by travertine platforms in the upper Salado in the vicinity of Paredes de Sigüenza and by quartzite gravels north of Imón and, in the lower Salado, north of Huérmeces.

---

[10] See Footnotes 6 and 7 on p. 115.

The 20 m Middle Terrace

Conglomerates, travertines, gravels, and silts some 20 to 50 m above the present streams define the Middle Terrace. This represents a complex and extended period of alluviation across the width of broad, gently concave valley floors. Conglomerates and some travertines formed in the stream beds, while other travertine deposits were laid down in ponded situations, and breccias accumulated in footslope areas or on upland surfaces. Gravel beds north of Huérmeces, as well as truncated platforms further downstream, record the Middle Terrace at 21 to 31 m in the lower Salado Basin.

The Low Terrace

In the upper Rio Salado, the Low Terrace is developed at 5 to 11 m above the campiña floor and recorded by tufa, caliche-coated gravel, and thin conglomeratic beds. Along the lower course of the river, this alluvial terrace is developed in the form of massive tufas and caliche-coated gravels at 14 to 18 m above the valley floor.

The Campiña Terrace

The campiña surface is defined as the present valley bottom incised 1 to 4 m by modern stream channels. Deposits in this terrace surface vary from silts and sands to gravels, and in the case of the lower Salado the Campiña Terrace is comprised, at least locally, of interbedded sandy silts, marls, and hydromorphic soil bands. This surface is coterminous with the Rio Henares Low Terrace at Baides.

The Rio Dulce Basin

The upper Dulce Basin

At Jodra del Pinar the Rio Dulce is joined by the Rio Saúca before cutting down through the limestones of the deformed Mesozoic-Tertiary contact. Upstream of this village the broad, flat-bottomed valleys grade in a gentle, concave fashion to a campiña margin 10 to 12 m above the present stream beds.

Noteworthy Pleistocene features of the upper Dulce are limited to a 1112 to 1128 m travertine platform that dips 2° toward the Rio Saúca, 1 km north of the village of Saúca (Fig. 46). The surface, consisting of as much as 80 cm of travertine, rises from 5 m to a maximum of 28 m above the river channel. The eroded rear segment of the platform extends to the edge of the valley and is buried by ancient colluvial deposits derived from the adjoining upland midslopes. As much as 3.5 m of unstratified but crudely sorted, coarse and fine layers of cemented detrital limestone comprise this slope breccia. Detrital litter derived from this breccia mantles a large portion of the downslope segment of

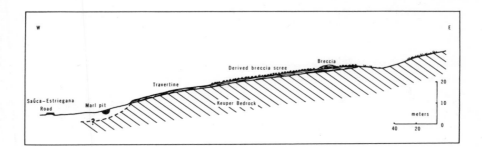

Fig. 46.--Transect of the upper Dulce Terrace. The marl pit, with tufa, is unrelated in time to the travertine platform. The former has yielded a Würm radiocarbon date.

the travertine platform. The breccia material is comparable to the slope deposits that have been described in association with the upper Henares and Salado Middle Terraces, so that the underlying travertine most probably relates to an older, High Terrace alluviation cycle, although direct stratigraphic analogies cannot be established because of the isolated occurrence of these deposits.

Embanked disconformably against the lower edge of this travertine platform is the following sequence of organic deposits, from bottom to top:

a) 80 cm of chalk white marl (Table 8, No. D703). The body of the sediment is powdery and unconsolidated; non-carbonate residue is dark brown (10 YR 4/3). Radiocarbon date, 25,300 B. P. ±750 (I-3544). Diffuse, wavy contact.

b) 12 cm of chalk white tufa (Table 8, No. D702). The bulbous tufa is weakly consolidated; non-carbonate residue is a reddish-brown (5YR 7/3) silt. The tufa grades laterally into an unconsolidated, very pale brown (10 YR 7/3) marl with crumb structure and occasional weakly consolidated masses. Irregular to wavy, diffuse contact.

c) 20 cm of very pale brown, silty clay (Table 8, No. D700A). This is the modern, loose, granular to fine crumb rendzina A-horizon that is developed on both the marl marl and tufa, and the travertine platform as well.

These younger deposits relate to a Würm alluviation phase that followed partial erosion of travertine deposits. This erosional period is of an undetermined age.

The campiña of the upper Dulce-Saúca drainage lies only 40 to 65 m below the general 1080 to 1140 m elevation of the uplands. This anomaly suggests that headwater dissection has been retarded, very probably due to the effective bedrock barrier of the Dulce gorge which has functioned as a high "temporary" base level. Within the gorge, medium-to-coarse, subrounded, quartzite and limestone conglomerates upstream of Pelegrina record an older floodplain some 30 m above the present river bed (1040 m). No strati-

TABLE 8

SEDIMENT CHARACTERISTICS OF THE RIO DULCE SYSTEM:
ESTRIEGANA, AND THE CAMPIÑA, LOW AND
MIDDLE TERRACES AT MANDAYONA

Sample Number	Strata	Color	Texture	Sorting	CaCO$_3$ % wgt	pH
Estriegana						
D703	Bed a	–	Marl	–	99.5	7.8
D702	Bed b	–	Marl	–	99.5	8.3
D700A	Bed c	10YR 7/4	Silty clay	Poor	82.7	7.7
Mandayona, Middle Terrace						
D901	Bed a	7.5YR 6/6	Sandy marl	Poor	52.4	8.0
D902	Bed b	5YR 6/6	Clayey coarse sand	Poor	51.6	7.6
D904	Bed c	5YR 5/8	Silty clay	Poor	26.8	7.4
D905	Bed d	7.5YR 6/6	Silty clay	Poor	33.3	7.6
D906	Bed e	7.5YR 7/4	Clayey sand	Poor	49.2	8.3
D907	Bed f	7.5YR 7/6	Sandy marl	Poor	76.5	8.6
D903	Bed g	5YR 6/6	Sandy clay	Moderate	63.7	8.0
Mandayona, Low Terrace						
D303	Level a	5YR 6/6	Coarse sand	Poor	23.3	7.8
D204	Level a	10YR 6.5/6	Coarse sand	Poor	22.5	7.6
D203	Level b	7.5YR 6/6	Silty sand	Poor	28.9	7.6
D301	Level c	5YR 6/6	Coarse sand	Poor	21.2	7.7
D201	Level c	10YR 6/4	Silty coarse sand	Poor	44.3	7.8
Mandayona, Campiña Terrace						
D501	Bed a	10YR 6/6	Coarse sand	Poor	25.3	8.0
D502	Bed b	10YR 5/2	Sandy marl	Moderate	50.3	7.3
D503	Bed c	7.5YR 3.5/0	Silty clay	Good	10.9	6.9
D504	Bed d	10YR 7/2	Marl	Good	82.8	7.5
D505	Bed e	5YR 5/6	Silty sand	Moderate	42.9	7.4

graphic significance can be attached to these materials because of their isolated occurrence in a gorge subject to intermittent downcutting through most of the Pleistocene.

The lower Dulce Basin

Downstream from Aragosa, the Dulce Basin is developed in the Tertiary materials of the Tajo Basin. The river valley widens and the gradient is reduced so that fluvial forms are better preserved along the valley margins. As in the case of the lower Salado Basin, however, Tertiary conglomerates confound unequivocal identification of some comparable Pleistocene materials. The abundant valley-side benches undoubtedly relate to several stages of development but, because of the dominance of structural controls and the absence of useful deposits, they are inadequate to establish Pleistocene terrace sequences. Alluvial terraces are well developed in the vicinity of Mandayona, however, where three terraces can be identified above the modern valley floor. These terraces, as well as correlative structurally preserved benches, are mapped in Figure 47; the relevant deposits are discussed below.

The High Terrace

At Mandayona, approximately 3 m of very well-cemented, light gray (10YR 6/2) travertines cap a bedrock terrace 40 m above the north bank of the Rio Dulce (+50 m above the river bed). Accumulations of this same sediment rest on red, silty sands of mid-Tertiary age on two adjacent platforms and are distinctly higher than the massive Middle Terrace tufa deposits across the river. The High Terrace also is recorded on the south side of the river 1.5 km further upstream, 45 m above the campiña (+50 above the river bed). This platform, cut across Tertiary materials, is mantled by medium-to-coarse grade, subrounded, limestone, dolomite and quartzite gravel and occasional pieces of tufa-travertine. The High Terrace also appears to be present 65 m above the Henares-Dulce confluence in the form of benches preserved on Tertiary conglomerates at Matillas (885 m).

The Middle Terrace

Due east of Mandayona, more than 4 m of well-cemented, pink (7.5YR 8/4) tufa overlie red Tertiary beds at 892 m. This terrace is perhaps the most prominent valley feature of the lower Rio Dulce, rising 28 m above the campiña surface (+35 m above the river bed) in the form of an inverted, truncated cone. It is capped by 25 cm of loose, stoney surface wash. Evidence of this same general terrace level is provided by tufa blocks 35 m above the river south bank, approximately 500 m farther upstream. Similar tufa beds also occur at Mandayona, 32 m above the river channel (892 m).

A road cut at the highway junction east of Mandayona has exposed a sequence of fluvial deposits graded to a floodplain level of the Middle Terrace. These may be summarized as follows from base to top (Fig. 48):

a) 250 cm of reddish-yellow, sandy marl (Table 8, No. D901). The body of the sediment is unconsolidated with medium crumb structure, and contains occasional lenses of pebbles, subrounded by solution. Basal gravels are coarse and poorly stratified. The deposits rest on Tertiary conglomerate. Gradual, irregular contact; disconformity.

b) 160 cm of coarse, subangular gravel in a reddish-yellow, clayey coarse sand (Table 8, No. D902). The limestone and quartzite gravels are crudely stratified throughout the body of the unconsolidated matrix; wavy bedding plans occur locally and basal pockets of the gravel have pyrolusite stains. Debris obscures contact.

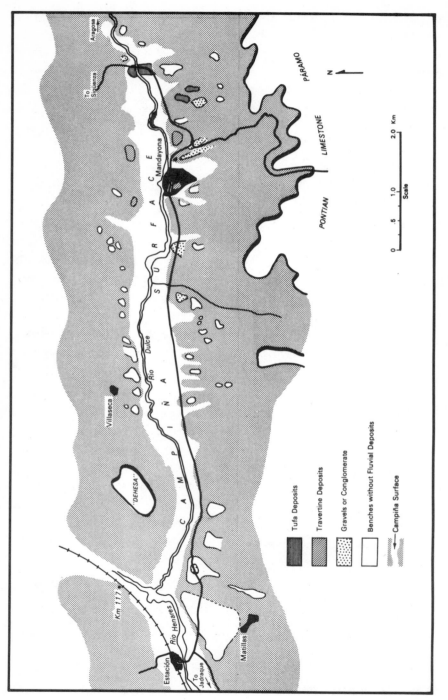

Fig. 47. --Pleistocene deposits of the lower Rio Dulce

Tufa Deposits

Travertine Deposits

Gravels or Conglomerate

Benches without Fluvial Deposits

Campiña Surface

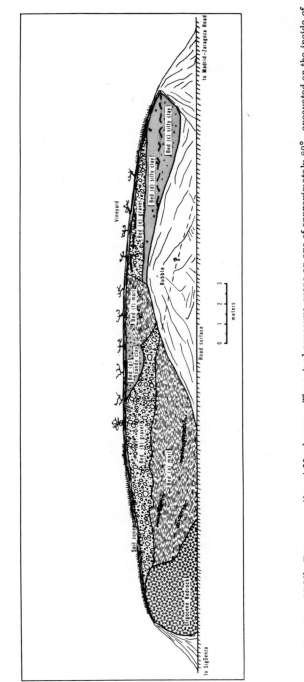

Fig. 48. --Middle Terrace section at Mandayona. The actual exposure spans an arc of approximately 80°, excavated on the inside of the Masegoso to Sigüenza road bend due north of Mandayona (Km 20.5). See text for description of deposits. The map scale is approximate due to distortion (no vertical exaggeration).

c) 150 cm of yellowish-red, silty clay (Table 8, No. D904). The sediment is unconsolidated, with medium angular blocky structure and occasional fine angular, limestone grit. Diffuse, irregular contact.

d) 40 cm of reddish-yellow, silty clay (Table 8, No. D905). The unconsolidated sediment is laced with chalk-white calcareous, root cast tubes or tunnels. Small fragments of shell can be detected locally, along with occasional limestone pebbles (<2 mm). Color and structure of the sediment vary somewhat vertically to red (2.5YR 4/6) in the upper 20 cm, and from medium coarse, subangular blocky-to-medium coarse angular blocky. Abrupt, smooth contact.

e) 80 cm of subangular gravel (Table 15) in pink, clayey, medium coarse sand (Table 8, No. D906). The well-stratified gravel is comprised of limestone, chert, dolomite and quartzite, in part cemented by calcareous crusts that incorporate portions of the otherwise unconsolidated sand matrix. The uppermost level of this sediment comprises finer, less gravelly alluvium. Clear, smooth contact; disconformity.

f) 130 cm of reddish-yellow, sandy marl (Table 8, No. D907). The weakly consolidated-to-weakly cemented sediment has coarse, crumb structure and occasional subrounded limestone pebbles and fragments derived from porous tufas. The uppermost 10-20 cm of these deposits are laced with small, white root drip. This sediment fills a channel trough cut into Beds (c) through (e). Gradual, smooth contact.

g) 90 cm of reddish-yellow, sandy clay (Table 8, D903). This unconsolidated deposit includes indurated, calcareous grit with pockets of colluvial wash. Small fragments of tufa material are also found in the body of this sediment.

The sequence is exposed between 895 and 897 m along the front of a smoothly convex spur that is inclined toward the Rio Dulce at 1° to 2°. The spur is capped by 40 to 50 cm of loose, yellowish-red (5YR 4/8) silty, medium sand soil-wash containing angular, limestone detritus. These deposits and the immediately adjacent Middle Terrace tufa platform (892 m) both rest on Tertiary bedrock (see Fig. 49). The similar basal elevations of the tufa and the sequence summarized above indicate that both phases of sedimentation were aggraded to the same local base level so that the two sites record different facies of Middle Terrace. The exact relationship is uncertain, unfortunately, because of subsequent dissection and modern road building, but the following events are recorded (from base to top):

a) Sedimentation of fine alluvium and gravels by streams eroding the bedrock of the valley floor (Bed a).

b) Erosional disconformity.

c) Alluviation of gravels by streams with relatively great competence and highly variable discharge, indicated by current bedding of coarse gravels (Bed b).

d) Renewed deposition of fine alluvium (Bed c).

e) Continued deposition by streams charged with carbonates leading to marl precipitation (Bed d).

f) Deposition of stream gravels, with some calcareous encrustation during drier phases (Bed e).

g) Erosional disconformity.

151

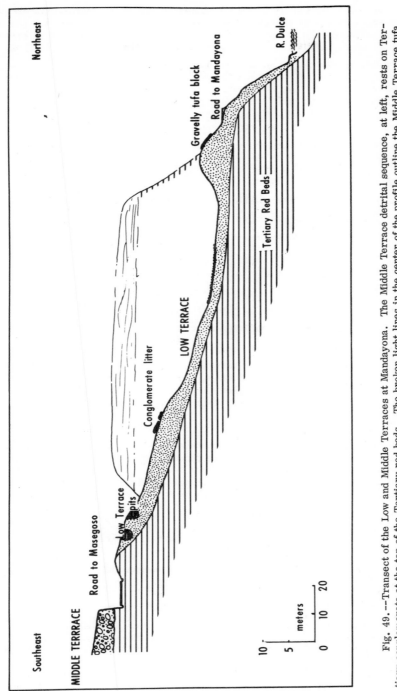

Fig. 49.—Transect of the Low and Middle Terraces at Mandayona. The Middle Terrace detrital sequence, at left, rests on Tertiary conglomerate at the top of the Tertiary red beds. The broken light lines in the center of the profile outline the Middle Terrace tufa platform immediately adjacent to (upstream) the transect. This feature is superimposed at its respective elevation, showing the similar elevations of the bedrock contact planes of the tufa and detrital sequence deposits (see text). Low Terrace deposits are described in the text.

h) Channel cutting and alluviation. Fragments of tufa debris included in these deposits suggests that accumulation of Middle Terrace tufas preceded this phase. The small hydraulic radius of this channel indicates a secondary drainage channel rather than a major valley floor stream; subsequent formation of root drip (Bed f).

i) Eventual spreading of colluvial wash (Bed g).

The Low Terrace

Deposits of the Low Terrace range from tufas or gravels in compacted but unce- mented sands, to gravels in cemented calcareous matrices. Below the Dulce gorge, these deposits are first recorded 6 to 8 m above the campiña east of the Mandayona-Sigüenza road.[11] Gravels derived from Tertiary conglomerate and weathered fragments of tufa and travertine veneer this Low Terrace platform. Similar gravels and an intact tufa cap 13 to 18 m above the campiña also record this terrace above Km 21.2 of the Mandayona-Sigüenza road, as does an 80 cm block of slumped tufa with basal gravels 7 m above the campiña near Km 8.7.

Gravels characterize the main body of the Low Terrace. Informative sections are exposed west of Mandayona (Km 7.2), immediately adjacent to and 9 m above the campiña, and by a pit excavated into the Low Terrace spur (see Fig. 49) east of Mandayona (Km 8.9). Taken in composite, these exposures indicate the following depositional sequence within the Low Terrace (from bottom to top):

a) Over 200 cm of subrounded, limestone gravel (Table 15) within a reddish-to- brownish-yellow, coarse sand (Table 8, No. D303, D204). The stratified medium- grade gravel is dispersed throughout the body of the consolidated sand. Subhori- zontal zones (3 cm) of carbonate enrichment occur in the lowermost portions of the sediment. Diffuse, wavy contact.

b) 22–100 cm of reddish-yellow, silty sand (Table 8, No. D203). The body of this sediment is weakly consolidated and moderately rich in carbonate material, and contains occasional pebbles. The uppermost approximately 20 cm vary from a strong brown (7.5YR 5/6), carbonate-rich, medium sand recorded at the pit east of Mandayona, to a pink (5YR 7.5/4), calcareous, fine sandy clay with tufa and shall fragments recorded in the exposure west of the village. Clear, wavy contact.

c) 15–40 cm of reddish-yellow or yellowish-brown, silty coarse sand (Table 8, No. D301, D201). The body of this sediment is loosely consolidated but with local calcretion; occasional, small angular or subangular limestone inclusions through- out.

The relationship between the Low Terrace exposures noted above is most clearly established by the basal sandy gravels at each site. These deposits are essentially

---

[11] The course and depth of the present river bed has been variously influenced and altered by damming, irrigation and other agricultural practices. As a result, relief along the river bank can vary from 2 to 15 m so that notation of Low Terrace elevations above the channel level or river bed would introduce a greater elevation range for this ter- race than is actually the case. Relative elevations of the Low Terrace are, therefore, cited in this section with respect to the less variable elevation of the campiña bottom.

identical. The overlying material records a horizontal facies change, since both sites are internally conformable. The darker color of Bed (b) at the pit exposure east of Mandayona suggests partial organic enrichment, perhaps the early development of a marginal flood-plain soil. This facies of Bed (b) is replaced by the marl-like sediment in the exposure west of Mandayona which is approximately 125 m closer to the modern Rio Dulce channel than the former site, and most probably was less marginal or even intra-floodplain in loca-tion with respect to its ancient river course. The uppermost site, east of Mandayona, is mantled by 15 cm of angular, Pontian limestone detritus; this veneer is less thick at the lower exposure west of Mandayona. In each case the material is representative of the val-ley detritus in Tertiary terrain.

Judging from the profile in Figure 49, alluvial fill of the Low Terrace may have reached a thickness of as much as 25 m. This fill has been remodeled by successively lower channel levels, as shown by the erosional shoulders developed in the deposits. Locally gravel aggradation was accompanied by ponding and precipitation of freshwater carbonates. This is indicated by a block of well-cemented, calcreted deposits resting against a flank of the Low Terrace at Km 8.7. The block, which slid part-way down the side of the terrace, comprises subrounded-to-rounded, dolomite and limestone gravel in a pinkish-white (7.5YR 8/2) sand with rare, small fragments of tufa.[12] The particular phase of Low Terrace alluviation recorded by these deposits very probably relates to the marly phase of Bed b recorded at the site west of Mandayona.

The Campiña Terrace

Depth of dissection of the Dulce campiña varies from 2 to 15 m. Locally a 2.5 to 3 m bench is developed along the active floodplain; at places, such as at the Dulce-Henares confluence, this level is mantled with quartzite gravels. A complete stratigraphic section of the campiña surface can be seen along the south bank of the river below the abandoned Parador de la Mina, 1 km downstream from the Mandayona-Sigüenza road bridge (Fig. 50). This section can be described as follows (base to top):

a) Over 200 cm of rounded gravels (Table 15) in brownish-yellow, coarse sand (Table 8, No. D501). The quartzite gravels are well stratified and locally cemented by the calcareous sand matrix. A cemented, 15 cm, sandy marl strata with shell fragments occurs within the body of the gravels, 150 cm from mean channel water level; radiocarbon date 19,700 B.P. ±400 (I-3543). Diffuse, wavy contact.

b) 30 cm of grayish-brown, sandy marl (Table 8, No. D502). The body of the con-solidated sediment contains occasional gravels. Diffuse, wavy contact.

---

[12]Similar conglomerates are found near Matillas, along Km 1.2 of the Estación de Matillas-Mandayona road. Here the conglomerates overlie Tertiary conglomerate so that Pleistocene origins are clearly demonstrated. As at Mandayona, erosional shoulders or benches also occur in the Low Terrace near the Henares-Dulce confluence.

Fig. 50.--Campiña Terrace at Mandayona. The exposure occurs on the left bank
of the Rio Dulce, 1 km downstream from the Mandayona-Sigüenza road bridge.

c) 70-100 cm of very dark gray, silty clay (Table 8, No. D503). Several zones can
be identified within the consolidated body of this organic material. The basal 30-
50 cm are of medium prismatic structure and only slightly humified; this grades
vertically into 5-10 cm of several laterally disappearing humic bands with columnar structure. The middle 20-25 cm of this bed are massively bedded with very
coarse, columnar structure. The topmost 15 cm comprise a consolidated, platy,
organic muck with charcoal fragments and light gray (5Y 5/1) root fills. Individual
peds retain ferruginous traces of oxidation. The organic component of the organic
muck has a radiocarbon age of 9,750 $\pm$160 (I-4594); the carbonate material has indicated a radiocarbon age of 12,570 $\pm$190 (I-4593) (see Appendix C). Abrupt, smooth
contact; disconformity.

d) 300 cm of light gray marl (Table 8, No. D504). The body of the marl is massively bedded, consolidated, and interdigited with bands of darker, organic sediment up to 10 cm thick and with some derived organic tufa material. Pockets of

tufa in the order of 1 m thick are common in this sediment which is exposed along both banks of the Rio Dulce. Radiocarbon age of the tufa material, obtained from a sample 1 km upstream, is 4020 B. P. ±110 (I-3112). Diffuse, wavy contact.

e) 135 cm of yellowish-red, silty sand (Table 8, No. D505). The consolidated sediment is calcareous and contains a small amount of limestone pebbles.

The Rio Dulce campiña sequence (Fig. 51) records a protracted phase of fluvial deposition in a riverine situation. Shifts of channel appear to be indicated by facies changes, although initial higher competence (basal gravels) was superseded by accumulation of sandy floodplain marls, suggesting more acquiescent alluviation, and ultimately by the intermittent development of humic floodplain soils with subsurface oxidation. Renewed sedimentation, in the form of marls and organic tufas, was also marked by seasonal groundwater-logging (see Fig. 51). Reddish colluvial wash, derived from the marls, clays and sands of the local Tertiary bedrock, comprises a thick mantle that overlies the exposures and much of the Campiña Terrace. Elsewhere thin veneers of gravel, or a thin soil profile of the rendzina type, with a brownish (10YR 5/4), A-horizon on the calcareous sediments, characterize the campiña surface. Strong analogies exist with this sequence and the Henares Low Terrace upstream from Baides, particularly in the succession of massive, well-stratified gravel, sandy marl, multi-hydromorphic soil horizons, and marl and tufa beds that are similar in each instance. In addition, the radiocarbon dates obtained from both sequences confirm these similarities. The erosional disconformities noted near Baides in all probability relate to local shifts of channel and overbank environments, rather than to major depositional breaks.

Summary

The three terraces of the Rio Dulce system are best preserved in the lower Dulce basin, particularly in the vicinity of Mandayona, where the following units can be recognized: 1) a 50 m High Terrace, 50 to 65 m above the river channel, characterized by massive travertine deposits; 2) a 35 m Middle Terrace, 35 m above the Dulce channel, indicated by massive tufas and by gravels; 3) a 12 m Low Terrace, 17 to 25 m above the modern river channel, recorded by tufas, sandy gravels and conglomerates; and 4) the Campiña Terrace, 3 to 11 m above the river, comprising marls and tufa with local gravel veneers. Precipitation of the Middle Terrace tufas was accompanied by fluvial sedimentation of sands and gravel. Aggradation of the Low Terrace was followed by an extended period of erosional remodeling of the Low Terrace floodplain, possibly synchronous with late Pleistocene deposition of gravels in the actual river channel. The campiña surface fill is of Holocene, and possibly in part of late Pleistocene age, as indicated by isotopic dating.

156

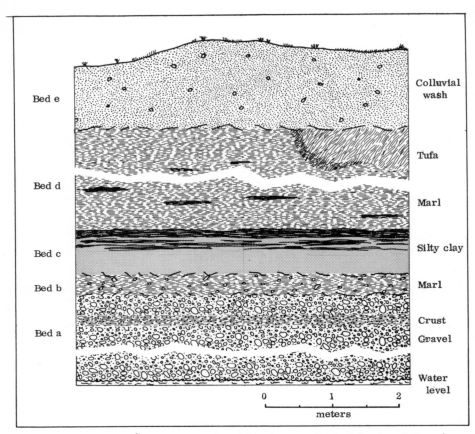

Fig. 51. Campiña Terrace section at Mandayona. See text for description of deposits. No vertical exaggeration.

Résumé and Comments

A suite of four terraces is well documented throughout the Alto Henares system. Deposits relating to the periods of terrace alluviation include travertines, tufas, marls, conglomerates, breccias and gravels, while intervals of chemical weathering are recorded by relict soils. These deposits are described in the preceding sections for the upper Henares, Salado, and Dulce systems and are mapped in Figures 19, 33, and 47. This sedimentological evidence, as well as altimetric and morphologic factors, was implemented to relate individual terrace features to counterparts elsewhere in the drainage system. The range of relative elevation of the remnants of a particular terrace reflects the fragmented occurrence of the features such that the field perspective was also vital to relate the terrace segments both to the ancient longitudinal profile and to the cross-section of the former floodplain of which it was a part.

The terrace stratigraphies discussed above were determined independently for each basin. These regional results have already been summarized in the sections dealing with each drainage system. The following synthesis can now be suggested for the Pleistocene and Holocene terrace stratigraphy of the Alto Henares Basin. Terrace elevations refer to the inferred elevation of the ancient terrace floodplain above the modern campiña while parenthetical elevations indicate the elevation ranges above modern river channels.

A 40 to 50 m High Terrace (28–80 m) is recorded in each basin by thick travertines. Locally, gravel mantles and breccias rest on such travertines.

An 18 to 35 m Middle Terrace (20–48 m) is marked by travertines, massive tufas, conglomerates and gravels that record an extended and complex period of alluviation further characterized by local interdigitations of marls, sands, or clays. Broad, smoothly concave valley cross-sections are suggested by lateral facies changes involving footslope screes, spring and pond precipitation of carbonates, and detrital aggradation of bed load deposits. Gravel fans were graded to this floodplain and breccias laterally include scree detritus and red soil sediments. The valley fills have a considerable thickness, frequently amounting to 20 or 30 m, with the interbedding of riverine, lacustrine and colluvial deposits. The waning phases of this period saw remodelling of the Middle Terrace deposits with local development of floodplain benches.

A 6 to 12 m Low Terrace (5–25 m) level is preserved by tufa deposits in each of the basins studied. Conglomeratic materials and calcrete-encrusted gravels are also characteristic. Soil formation climaxed this particular cycle.

The present valley floor, or Campiña Terrace, is found at 5 to 7 m above the stream channels. Tufa, marl, and gravelly sands comprise this surface on which halomorphic or humic alluvial soils can be observed locally. Radiocarbon dating of the tufas indicates a mid-Holocene age for the Campiña Terrace.

These fluvial stages establish the broad lines of the Pleistocene succession for the Alto Henares. Alluvial terraces record periods during which, for one reason or another, fluvial aggradation was dominant. These phases were punctuated by intervals of downcutting accompanied or followed by general denudation. Both the tectonic stability characteristic of this area throughout most--if not all--of the Pleistocene, as well as the nature of the Pleistocene sediments point to climatic fluctuations as the triggering mechanism of these cycles of alluviation and dissection. The following sequence of Pleistocene and Holocene events is indicated by the terrace suite and its related deposits:

a) Alluviation of the High Terrace floodplains in valley bottoms situated below the youngest "D" planation surface (see Chap. V).

b) Dissection of the High Terrace floodplains.

c) A protracted and complex period of Middle Terrace alluviation with marginal precipitation of freshwater carbonates. Several major phases of aggradation with

intermittent erosion are indicated. Of environmental interest is the later development of valley entrant fans, graded to the general level of the valley bottom, as
well as a variety of upland, slope, and valley-side screes or breccias.

d) Periods of calcretion and intense chemical weathering, exact temporal relations
uncertain.

e) Renewed channel downcutting with dissection of the Middle Terrace.

f) Alluviation of the Low Terrace floodplains in comparatively narrow valleys,
accompanied by tufa precipitation.

g) Period of soil formation.

h) Dissection of the Low Terrace floodplains.

i) Cutting of modern stream channels with local lateral erosion and the occurrence
of fluvial benches within the channels.

CHAPTER VII

INTERPRETATION OF PLEISTOCENE SEDIMENTS

FROM THE ALTO HENARES

The terrace suite of the Alto Henares may be used to relate the sedimentological record to a broad Pleistocene framework while aspects of the deposits, as well as the isotopic dates obtained, furnish a basis for more detailed interpretations of glacial and interglacial conditions. This chapter focuses upon the stratigraphic implications of the deposits of the Alto Henares terraces and briefly summarizes additional information recorded elsewhere in the Hesperian Meseta.

Terrace Origins

Convolutions apparent in the Middle Terrace deposits illustrate the interrelationship of the hydraulic cut-fill cycle with the glacial-interglacial climatic rhythm of the Pleistocene. Since cold climate conditions and intensive soil frost are requisite for development of such features, the associated fluvial deposits were aggraded prior to (or during) the "periglacial" conditions recorded in the Middle Terrace type sequence at Riba de Santiuste. Comparable features are not evident in deposits of the High, Low, or Campiña Terraces, but these formations do not provide equally complete and informative sequences. The sedimentological indications are, therefore, that the upper courses of Hesperian rivers aggraded their channels during colder, glacial stages of the Pleistocene, dissecting their channel fills during the waning glacial phases or during more temperate interglacial periods.[1] Consequently, in this instance the nature of the fluvial deposits themselves resolves controversy over the particular response of the hydraulic regimen to Pleistocene climatic change.

This conclusion is strengthened by stream behavior under the present non-glacial conditions. As noted in Chapter III, channel activity of major streams in the upper basins consists essentially of the reworking of older fill. The seasonal periods of increased discharge promote greater proportions of suspended load and local channel scour of older

---

[1]It may be noted that this locally derived interpretation conforms to the model for most central European rivers during the Pleistocene (Schaefer, 1950).

alluvia. The middle Henares course below Jadraque, as well as the lower Rio Henares downstream from Humanes, relates the same overall tendency; throughout these river segments the stream bed comprises older gravels reworked by the modern river channel.

The conglomerates and breccias associated with the Middle Terrace alluviation also relate to cooler and/or moister phases of the Pleistocene (see Chap. VI). The detrital component of the slope breccias comprises solifluctoidal scree intermixed with fine, colluvial sediment--later cemented by carbonate-charged runoff, possibly originating in part from snow meltwaters. The Middle Terrace conglomerates are channel-bed counterparts of the slope breccias.

The laminated travertines and massive tufas of the Hesperian Meseta are of the sort commonly attributed to non-glacial conditions (Slack, 1967). Warm or temperate climates favor greater evaporation rates, to produce the supersaturation of carbon dioxide that is requisite for carbonate precipitation. Accretion of these sediments may have continued during colder phases of the Pleistocene as well, but for a rather different reason. Increased rates of carbonation (solution by carbonic acid) during these colder phases would also have produced carbon dioxide supersaturation and extensive precipitation of calcium carbonate.[2] The essentially non-detrital composition of the accretions in the study area implies a complete vegetation mat that effectively inhibited soil stripping or colluvial washing. Such protective conditions may have pertained to both glacial and non-glacial phases but in any event the implication is one of morphogenesis under conditions at least as moist as today if such a vegetative cover was to have prevailed. The actual conditions of travertine accretion remain conjectural, but of interest and relevance is the fact that deposits of this generic type have not been recorded forming in the Hesperian Meseta under contemporary non-glacial conditions.

In summary, the morphogenetic implications of the clastic and calcium carbonate sediment types of the study area tend to confirm that fluvial deposits were aggraded during Pleistocene glacial phases, followed by interglacial fill dissection that stranded the fluvial deposits as terrace segments.

Implications of the Terrace Sequence

The detailed stratigraphic record described in previous sections invites paleo-environmental and climatic inferences beyond the general glacial/non-glacial distinctions outlined above. In particular, the clear record of the Middle Terrace at Riba de Santiuste and the sequences of the Low and Campiña Terraces, with radiocarbon dates, encourage

---

[2]Algal discoloration of travertine accumulations in the study area is not prominent, suggesting further that carbonate precipitation was primarily the result of inorganic processes.

more elaborate inferences with regard to paleosettings. Caution is due, however, since
without palynological and/or paleontological information, paleoenvironmental reconstructions per se are not appropriate. Furthermore, morphogenetic environments indicated by
the types of deposits encountered in the Alto Henares are not exclusive to individual times
within the Pleistocene (see Table 9) and, for the most part, the various exposures permit
only a sedimentological sequence to be established, in some instances along with general
conclusions regarding Pleistocene environments. The appropriateness of such restraint
is emphasized by the uncertainty surrounding the actual nature of Mediterranean paleo-
climates during the glacials. Whereas the non-glaciated regions of the northern Mediterranean Basin are widely held to have been cool and humid during the glacial stages,
recently published studies of dated pollen sequences indicate that these areas may, in fact,
have been decidedly arid during glacial climatic extremes (see discussion in Bonatti, 1966).
Consequently, geomorphic inferences based upon the Alto Henares terrace sequence considered below relate solely to the morphogenetic implications of the deposits described in
the previous chapter.

The 40-50 m High Terrace

The oldest Pleistocene floodplain level of the Alto Henares is recorded by thick travertine accumulations in the three basins studied. There is no type-site per se for the
High Terrace since no informative sequences underlie the flowstones while the gravels
recorded elsewhere are surficial deposits without stratigraphic context.

The infrequent occurrence of High Terrace deposits is an indication of their remote
origin, since only a trace of this alluviation cycle has survived. Great antiquity is also
implied by the altimetric situation of the High Terrace directly below the Plio-Pleistocene
"D" erosional surface, but several tens of meters above the post-Pleistocene campiña
floor. In contrast to Middle Terrace platforms, massive travertine slump blocks litter
High Terrace midslopes and point to protracted and/or recurrent periods of undercutting
throughout a major time span of the Pleistocene. Little more can be added to the morphogenetic implications of these accretionary deposits beyond that discussed in Chapter VI.

The 18-35 m Middle Terrace

The Middle Terrace alluviation cycle is the most thoroughly documented fluvial
sequence in the Alto Henares. The characteristic interbedding of clastic and calcium carbonate facies reoccurs with local facies variation near Horna, at Riotovi, near Mandayona,
and most extensively at the Riba de Santiuste type-site.

The great overall thickness of Middle Terrace fill at these locations portrays an
extended period of deposition. This time span experienced repeated variations in stream

TABLE 9

STRATIGRAPHY OF SELECTED WESTERN MEDITERRANEAN DEPOSITS

	Central Spain	Mediterranean Spain	Balearics	Mediterranean France
Würm Interstadials	Paleosols (Riba, 1957; Vaudour, 1969)	Peat (Solé, 1962) Terra fusca, calc-braunerde (Brunnacker & Ložek, 1969)		Red soil (Bonifay, 1957, 1959) Colluvial silts (Bonifay, 1959)
Würm		Crusts (Rutte, 1958) Travertine (Klinge, 1958) Colluvial silts (Butzer, 1964a)	Colluvial silts (Butzer & Cuerda, 1962) Travertine (Butzer, 1964a)	
Riss/Würm	Soil (Riba, 1957; Vaudour, 1969)	Red soil (Klinge, 1958) Braunlehm (Butzer, 1964a)	Terra rossa (Butzer & Cuerda, 1962)	Red soil (Bonifay, 1957, 1959) Travertine (Bonifay, 1962)
Riss		Travertine (Rutte, 1958; Klinge, 1958; Butzer, 1964a)	Colluvial silts (Solé, 1962; Butzer & Cuerda, 1962) Travertine (Solé, 1962)	Travertine (Bonifay, 1962)
Mindel/Riss	Red soil (Riba, 1957; Butzer, 1965; Vaudour, 1969)	Rotlehm (Butzer, 1964a)	Terra rossa (Butzer & Cuerda, 1962)	Red soil (Bonifay, 1957)
Mindel	Colluvial silts (Butzer, 1965)	Colluvial silts (Solé, 1962)	Colluvial silts (Butzer & Cuerda, 1962)	
Pre-Mindel	Early Pleistocene red soil (Vaudour, 1969)	Mio-Pliocene red soil (Klinge, 1958) Plio-Villafranchian rotlehm (Butzer, 1964a)		

discharge as well as several intervals during which erosion predominated. In addition to variability of discharge, as indicated by the size-range of clastic sediments (clays to gravels), fluctuation in the water table also is indicated by groundwater horizons noted in several of the facies. On the basis of these factors and the calcareous crusts, it may be inferred that the moisture balance varied markedly during the Middle Terrace alluviation hemicycle, as also must have temperature as shown by soil-frost convolutions in several of the beds. Neither the degree of change nor the behavior of the temperature and precipitation patterns can be determined on the basis of the sedimentological data, nor can the vegetation be inferred without palynological data. Of note in the latter instance, however, is the fact that the marly beds are conspicuous because of their homogeneity, being almost entirely without clastic material of mineral or organic sorts. This suggests limited or negligible subaerial denudation, perhaps as a consequence of an effective vegetation mat.

The overall morphogenetic sequence recorded at the Riba de Santiuste type-site, as well as at the additional Middle Terrace exposures, have already been outlined in Chapter VI. In light of the foregoing comments, the Middle Terrace sequence may be summarized as follows (based on the Inclined Sequence at Riba de Santiuste):

Alluviation phase 1: Initial period of inhibited discharge and fluctuating water table (Bed a marl), followed by a period of increasingly more turbulent stream discharge (Beds b and c). The terminal aggradation was marked by reduced competence (Bed d).

Erosional phase 1

Alluviation phase 2: Introduced by high stream competence (Bed e gravel), superceded by increasingly moderate discharge (Bed f sand and marl) under cold-climate conditions.

Erosional phase 2

Alluviation phase 3: Initially great stream competence (Bed g gravels) under cold conditions, subsequently becoming more moderate (Bed h sand with gravel) and dry (calcrete crust).

Erosional phase 3 (?)

Alluviation phase 4: Introduced by a period of inhibited runoff with a high groundwater table (oxidation), punctuated by periods of more turbulent flow (current-bedded medium-to-fine sand, Bed i). Subsequent floodplain conditions were marked by slow, intermittent accumulation of carbonates (Bed j marl with oxidation) followed by another phase of turbulent discharge (current bedding of Bed k sands).

Erosional phase 4

Alluviation phase 5: Begun under conditions of moderate discharge but becoming more turbulent (current bedding), with intensive soil-frost (convolutions) and a relatively dry climate (calcrete crust) (Bed l gravel). A period of less turbulent discharge culminated this phase under less cool conditions.

Erosional phase 5 (of uncertain duration and extent).

Alluviation phase 6: Generally high competence of discharge, alluviating the Middle Terrace conglomerate cap with subsequent calcretion, probably under drier conditions.

It is apparent from the foregoing succession that each of the erosional disconformities was preceded by a period of relatively quiet stream flow and succeeded by a period of abruptly increased competence or turbulence. It is uncertain whether this approximation of graded bedding reflects on aperiodic shifts of channel location or on long-term changes of geomorphic steady state--and ultimately of climate. The repeated development of calcrete crusts or of soil-frost phenomena favors the latter interpretation.[3]

The interbedding of the clastic fluvial and intrastream or limnic-type calcareous sediments is sufficiently distinctive within the study area, and the Hesperian Meseta in general, to allow such exposures to be related unequivocally to the Middle Terrace sequence. As already noted, there are no High Terrace exposures in any way similar; although the two Low Terrace sequences recorded below the present campiña surface (at Baides and near Mandayona) are characterized by both calcareous and clastic deposits, these sites contrast with the Middle Terrace deposits in sediment properties and by their lack of cold climate indicators. On the basis of the internal arrangement of Middle Terrace clastic-marl facies rhythm, the indicators of cold climate, and the disconformities, certain similarities exist among the several Alto Henares Middle Terrace deposits and the Torralba Formation recognized in the adjoining drainage basin by Butzer (1965). Direct facies correlations are not possible nor is every aspect of each sequence directly relatable to that of each of the others. The correlations suggested in Table 10 therefore are in part inferential. The Riba de Santiuste-Torralba Formation analogies may be considered fairly reasonable, however, because of the completeness and detail of the sedimentary sequence recorded in each case.

The thick travertine accumulations of the Alto Henares Middle Terrace record local variation of the protracted valley floor aggradation preserved at Riba de Santiuste. Where exposed, deposits underlying the travertine indicate marginal or intrastream situations; frequently the basal travertine rests on carbonate-rich clays, silts and fine sands. Occasional oxidation stains and micro-caliche bands suggest alternating waterlogging and dehydration during or immediately subsequent to sedimentation by slowly moving water. Reddish-brown hues may also derive in part from stripped soils or soil-derived sediments; clear evidence of surface weathering is not recorded.

The original situation of the thick travertine accretions is suggested in part by a modern analogy. Near the El Molar travertine platform a spring, which serves as the laundry site for the village of Mojares, is depositing a thin film of calcareous sediment.

---

[3] Of some interest is the fact that no evidence of paleosol development is indicated throughout this Middle Terrace sequence and is only suggested elsewhere within the study area. This is consistent with the interpretation above wherein the morphostatic periods of soil formation may well have been dry and cold as well.

TABLE 10

SUGGESTED CORRELATION OF ALTO HENARES MIDDLE TERRACE DEPOSITS WITH THE TORRALBA FORMATION

Torralba Formation (Butzer, 1965)	Riba de Santiuste Middle Terrace		Riotovi Middle Terrace		Horna Middle Terrace	
	Bed		Bed		Bed	
Vd "D-gravel" (cryoclastic)		Conglomerate cap				
Vc Fine red alluvium	m	Silty sand			l	Calcreted silt
Vb "C-gravel" (cryoclastic) Va Marly sand	l	Gravels with crust	g	Gravel with basal convolutions & capping crust	k j i	Silty conglomerate Gravelly sand Conglomerate
∿ disconformity ∿						
IVb Gray marl	k j	Marl with sand & convolutions Marl & convolutions	f	Yellow marl & tufa	h	Silt with calcareous crusts
IVa Clastic gray marl	i	Marl with sand	e	Brown marl with gravels	g	Brown marl
∿ disconformity ∿ — ? ∿			∿ — ? ∿			
IIIb "B-gravel"	h	Pebbly sand	d	Sandy marl with basal involutions	f e d	Conglomerate and marl Marl
IIIa Gray colluvium with solifluction	g	Gravel with basal convolutions				
∿ disconformity ∿						
IId Brown marl	f	Sand & marl with convolutions			c	Calcreted marl
IIc Gray colluvium with involutions			c b	Marl with sand & gravel Marl with sand	b a	Marl Calcreted marl
∿disconformity ∿						
IIb "A-gravel" (cryoclastic) IIa Gray silt with involutions	e	Gravel				
∿disconformity ∿						
I Red colluvium (cryoclastic)	d c b a	Sandy marl Gravel Pebbly sand Marl	a	Subrounded gravel		

The origin of the spring is located near the upland limestone-Keuper shale contact, approximately 60 m above the present course of the Rio Henares which is located some 600 m away. Such a valley-side setting may represent the origins of the travertine terraces which formed from the combined sedimentation of numerous valley-margin springs. A location adjacent to or within an active stream channel would account for the interbedding of carbonate and clastic facies that characterize the Middle Terrace, where the spring sites were intermittently subject to alluviation by a laterally migrating channel. Travertine flowstones occur in altimetric correlation with the Middle Terrace conglomerates where such deposits are in proximity and since a conglomerate caps the Middle Terrace sequence at Riba de Santiuste, it appears that travertines and conglomerates (and breccia) represent the waning period of Middle Terrace sedimentation.

Of final note are the deposits graded to the Middle Terrace of the lower Dulce, near Mandayona. The exact relationship of these beds to the adjacent dense tufa accumulations cannot be demonstrated because of subsequent dissection and modern road building. Both sites relate, however, to the same general alluviation cycle because both rest on bedrock and are aggraded to the same immediate base level. These deposits record several phases of sedimentation which reflect an interplay of stream channel deposits and lateral entrant fans and channels (arroyos?), punctuated by erosional disconformities and perhaps soil development.

A comparatively long time intervened between terminal alluviation of the Middle Terrace and the initial phases of Low Terrace development. This is indicated by the excavation and incision of Middle Terrace fill in the form of bench-like features still preserved in the Arroyo de la Calera, at Mandayona, and at many additional locales where Low Terrace deposits are in proximity to the Middle Terrace.

The 6-12 m Low Terrace and Campiña Terrace

Deposits associated with the Low Terrace floodplain include conglomerates, tufas or tufa-marls occurring throughout the upper basins of the Salado, Henares, and Dulce. Upstream, Low Terrace remnants retain distinct morphologic definition, clearly offset below the older Middle Terrace, yet ranging from 9 to 17 m above the present river channels. The calcareous sediments are of particular interest because of the Würm radiocarbon ages obtained. The Horna tufa date (I-3111) indicates that Low Terrace deposits there appear to date from the Würm Interpleniglacial while the upper Dulce marl (I-3544) is somewhat younger, dating from the Upper Pleniglacial. Because of the close agreement of these dates, the Low Terrace of the upper basins of the study area is considered to date from Würm times. A warp-type paleosol (braunlehm vega) occurs in the Low Terrace of the Arroyo de la Calera, east of Sigüenza. Whereas the terrace surface is now

graded to the modern campiña, the paleosol horizon grades to the older and higher Low
Terrace floodplain that would have been contiguous with the tufa deposits preserved farther upstream.

In the upper Salado, Henares and Dulce basins the extensive valley floor campiñas
are younger than the Low Terrace deposits and possibly record some post-Pleistocene
alluviation. Modern streams are dissecting these surfaces and reworking older deposits.
Downstream from the Mesozoic-Tertiary contact interpretation of the Low Terrace-
Campiña Terrace relationship is complicated by contrasting altimetric, morphologic and
sedimentologic properties reflecting on a different geologic province. Other aspects of
these differences also result from the knickpoints in the river longitudinal profiles that
mark the geologic contact.

The relationship of the detailed Baides and Mandayona sequences as supported by
the isotopic chronology is shown in Table 11. Facies differences are considered to reflect
differences in local setting and erosional breaks record local events within the valley bottom. This procedure equates sedimentation of gravel and marl, and the development of
calcrete crust in the lower Dulce (Beds a and b) with marl and sand (Bed f) of the Rio
Henares. The subsequent morphostatic phase allowed a floodplain soil to develop, but
whereas this particular phase was comparatively protracted at the Dulce site, as shown
by the thickness of deposits, stronger structure, and more intensely developed organic
horizon, the weaker soil horizon at Baides indicates interruption by local erosion, with
floodplain sedimentation (Bed g) and some subsequent erosion. Terminal alluviation at
each site comprised marl and tufa materials in ponded or swamp settings. Along the
lower Rio Dulce this entire sequence is buried under sandy colluvial wash and although not
recorded in the stratigraphic sequence described at Baides, this same colluvial material
is present farther upslope above the Baides type-site.

Since the chronology of the Baides and Mandayona sequences spans Upper Pleni-
glacial-to-Holocene time, the relationship of the morphologic and altimetric Low Terrace
of the Tertiary geologic province appears distinctive from its upper basin counterpart. In
the latter area, the Low Terrace has been shown to date from mid- to late Würm time,
but since equivalent deposits downstream are recorded within the Campiña Terrace, the
Low Terrace of the lower Salado and Dulce is presumably older. Low Terrace deposits
comprising sandy limestone gravel beds with calcareous encrustations have been described
for both of these basins and, in the Lower Dulce, a great overall thickness of fill (ca.
25 m) with occasional remnant tufa blocks further characterizes Low Terrace alluviation.
Low or moderate elevation of these features above the Lower Dulce campiña surface, generally poor morphologic definition, and comparatively scarce occurrence of deposits are
characteristic of this older terrace. At this point definite relationships cannot be demon-

TABLE 11

STRATIGRAPHY OF WÜRM AND HOLOCENE DEPOSITS

	Rio Henares near Baides			Rio Dulce near Mandayona	
Bed h	Colluvial wash		Bed e	Colluvial wash	
	Tufa	6560±130 BP	Bed d	Tufa/marl	4020±110 BP
	Erosion				
Bed g	Marl				
	Erosion			Erosion	
Bed f	{ Organic marl Shelly sands Marl	19,450±350 BP	Bed c	Organic clay	9750±160 BP (organics) 12,520±190 BP (carbonates)
Bed e	{ Organic silt Silty sand		Bed b	Sandy marl	
			Bed a	Gravels	19,700±400 BP (crust)
	Erosion				
Bed d	Marl and gravels				
Bed c	Sandy marl				
Bed b	Pebbly sand				
Bed a	Clayey sand				

strated but the Low Terrace tufas of the Dulce are considered to relate to an earlier Würm period, while the clastic sediments within the body of the Low Terrace are regarded as an older fill.

Stratigraphic Correlation of the Terrace Sequence

None of the widespread materials discussed in the previous chapter are exclusive to a particular stage of the Pleistocene, so that they serve as time-stratigraphic indicators in a qualified way only. This is demonstrated most clearly by the multiple occurrence of specific sediment types in different terraces of the study area and during different Pleistocene stages within the western Mediterranean Basin (see Table 9). Travertines, for example, occur within the context of the High and Middle terraces and fluvial conglomerates within Middle and Low terrace contexts. In addition, marls and tufas are shown by isotopic dating to relate to Pleistocene and Holocene deposition, while paleosol sediments or horizons have been noted in the Middle and Low terraces respectively. Other studies which also demonstrate the temporally-reoccurrent aspect of these materials may be cited for the western Mediterranean Basin, as shown by the brief résumé in Table 9. Time-stratigraphic implications of the upper Henares terrace sequence must be derived, therefore, from the alternative of relating major fluvial terrace levels to the long-term glacial-interglacial rhythm of the Pleistocene. Such a procedure is enhanced here because of the absolute dates now available for the Campiña and Low Terraces.

Pleistocene Terraces from Areas Adjacent
to the Alto Henares

The middle Henares

A suite of fluvial terraces is prominently developed along the Rio Henares downstream from the study area. These platforms occur above the right bank of the river between Jadraque and Espinosa de Henares (Fig. 52). Pleistocene implications were first noted by Schwenzner (1937:50-53) and partial mapping has been done in a rather general way by de la Concha (1963). A more precise definition of the middle Henares terraces, based on field reconnaissance, is shown in Figure 53.

The campiña of the middle Henares, 5-8 m above the river, is referred to here as the Lower Low Terrace (LT 2), offset 2 to 5 m above the modern floodplain. The broad, nearly flat expanse of this level is entirely cultivated through this segment of the Rio Henares and is distinct from abandoned gravel and sand meander scars within the river channel itself; these younger features are locally stabilized by shrub and woodland vegetation or reclaimed for huerta cultivation.

Fig. 52. --Terraces near Jadraque (view to the northwest). The highest platform level comprises the C surface of Pontian limestone below which several of the terraces noted in the text are preserved. The Cordillera Central are in the distance.

A Higher Low Terrace (LT 1) is found 15 to 20 m above the river mean water level, but it may coalesce laterally with LT 2. This higher level is best recorded above the Rio Henares right bank, downstream from Jadraque at the Brihuega-Atienza road bridge where more than 2 m of crudely stratified, subrounded-to-rounded, coarse limestone and quartz- ite gravel with occasional calcareous encrustations are exposed. It is not certain whether this fill relates to Upper Low Terrace alluviation or represents a remodeled Middle Ter- race deposit.

The several Middle and High terrace levels are best developed and recorded between the confluences of the Rio Bornova and Rio Cañamares with the Henares. The

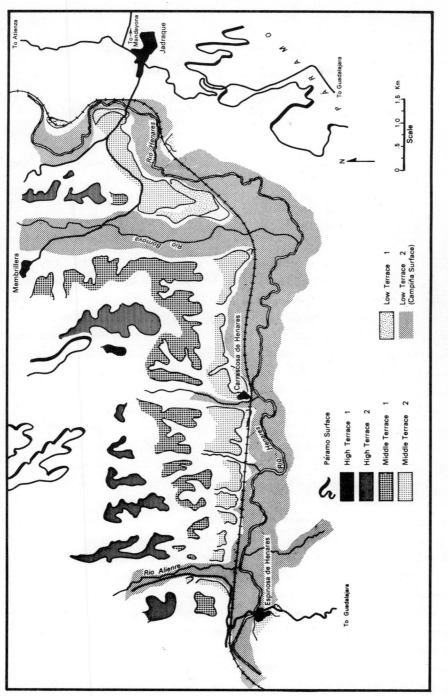

Fig. 53. --Middle Henares terraces in the Jadraque area

Brihuega-Atienza road traverses each of these levels so that the terrace deposits may be located in relation to this route:

>Lower Middle Terrace (MT 2, 25-30 m). Arroyo incision at Km 31.1 exposes a 10 m sequence of: a) 3 m basal silt; b) 2 m lenses of coarse, subrounded, quartzite gravels, in part calcreted to form a conglomerate; c) 3 m, upper silt; and d) 1 m sands. Calcrete horizons may grade laterally into crusts. Quartzite cobbles and boulders, derived from local masses (1.3 m) of conglomerate (Pleistocene?) exposed along the road to Membrillera, litter this surface. The broad platform between the Cañamares and Bornova at 800 to 815 m relates to this terrace level. More than 2 m of similar gravels are exposed at Km 32.8 of the Brihuega-Atienza road at 820 m.

>Upper Middle Terrace (MT 1, 70-90 m). Coarse, quartzite gravels littering the ground record this level at 860 to 880 m between Km 33.5 and 34.

>Lower High Terrace (HT 2, 110-115 m). Recorded by crudely stratified, coarse quartzite gravel at Km 35.5 at 900 m.

>Upper High Terrace (HT 1, 130-135 m). At least 5 m of unstratified subangular quartzite gravel in a loose, coarse sand matrix characterize this terrace between Km 35.9 and 37.5 (920-925 m). This level is altimetrically lower than and presumed younger than the erosional surface west of Membrillera (940-990 m).

The relationship of the middle Henares terraces to the upper Henares terraces identified in this study is not easily established. This results from the fact that there are no detailed or dated stratigraphic sequences along the middle Henares course. Consequently, at this point interrelationships have been established by tracing the morphologic entities downstream by field reconnaissance, aided by aerial photography. It has been noted that at Baides the flat valley bottom of the lower Salado conjoins with its Rio Henares counterpart; these surfaces have been described as Campiña Terraces of the respective drainage systems, as a rule rarely attaining more than 5 m elevation above the modern floodplains. In the vicinity of Baides, however, this surface is more deeply incised so that it is preserved in the form of a low terrace approximately 10 m or more above the river, and the campiña surface proper becomes a distinctly lower feature in this segment of the river. This low terrace level may be followed discontinuously to Jadraque where it appears to merge with the 15 to 20 m Upper Low Terrace and where, since the valley bottom has less lateral constriction, the meandering river course has additionally developed the 5 to 8 m Lower Low Terrace. If this correlation is accepted, the upper Henares terraces can be readily related to the well-developed fluvial platforms of the middle Henares.

Schwenzner's treatment of the Rio Henares terraces is noteworthy and deserves comment, if only because it is without precedence. The following summary discussion is based in part on the text (Schwenzner, 1937:50-53) and in part on Profile 7 of his monograph. Upstream from Baides, Schwenzner identified only two terrace levels in the upper Henares; these were correlated with two high terrace levels of the middle Henares in the vicinity of Jadraque ("OHT" and "UHT"). Knickpoints were depicted in the longitudinal profiles at the Mesozoic-Tertiary geologic contact near Baides such that only these highest terrace floodplains developed upstream from this point. Because only two terraces were recognized in the upper Henares, it appeared necessary that platform remnants there be assigned to one of these two levels, a procedure which confuses the correlation of Schwenzner's terraces with those identified in this study. Downstream from Baides, the High Terrace levels are noted along both sides of the Henares, forming platforms 80 to 140 m above the river.[4]

According to Schwenzner, the two Middle Terrace levels identified near Jadraque are recorded in the lower Salado basin. The Upper Middle Terrace ("OMT") may be found 25 to 30 m above the valley bottom, and the Lower Middle Terrace ("UMT") that occurs as a 30 to 40 m platform near Jadraque, becomes a 10 to 15 m "floodplain" (Talaue) or campiña terrace in the lower Salado. The relative elevations of these terraces are lower than near Jadraque, presumably because of slower bedrock cutting by the Rio Salado. Neither of the low terraces that Schwenzner identified in the lower Henares (downstream from Humanes) is recorded in the upper Henares (upstream from the Cañamares-Henares confluence) and although these levels are shown in Profile 7 to occur in the middle Henares, there is no text reference to their presence at Humanes.

The relationship of the middle Henares terraces as identified by Schwenzner to the middle Henares levels identified in this study is shown in Table 12. Aspects of this correlation are necessarily inferential because of vagueness in Schwenzner's account; he did not map the terraces nor were deposits described. In only two cases is the text detailed enough to allow positive location of features and establishment of relations to the respective counterparts as mapped in Figure 53. The remaining terrace relationships indicated in Table 12 are possible but speculative.

It is relevant to mention that the same difficulty applies in identifying the terrace levels mentioned by Schwenzner in the lower Henares and in relating these to their respective counterparts as outlined in the following section of this study. At Humanes, for example, seven distinct levels are shown by Schwenzner in Profile 7, but only three of these are mentioned in the text, and just two are sufficiently described to permit identification in the field. Furthermore, it is by no means certain that Schwenzner's longitudinal profile reconstructions are accurate correlations of horizontally disjunct terrace remnants.[5]

---

[4]Structural origin of these platforms is not precluded since, with only one exception, fluvial deposits were not noted. The exception pertains to gravels recorded on top of the lower high terrace ("UHT") at 80 to 90 m between Baides and Viana de Jadraque. The occurrence of these "gravels" is indeed sparse and the platforms themselves are developed on or near Tertiary conglomerate so that unequivocal citation as "terraces" is not assured.

[5]An indication of this again is provided in the treatment of the terraces at Humanes. Text references are to: a 15 to 20 m Upper Low Terrace on which the railroad station is

TABLE 12

TERRACE LEVELS OF THE ALTO AND MIDDLE HENARES

Alto Henares	Middle Henares	
This Study	This Study	According to Schwenzner (1937)
Campiña surface Terrace: +5-7 m	Lower Low Terrace (LT 2): 5-8 m	
Low Terrace: +6-12 m	Upper Low Terrace (LT 1): 15-20 m	"ONT" at 15-20 m. Well developed along both river banks from the Cañamares to the Jarama confluence; "Estacion" level at Humanes
Middle Terrace: 18-35 m	Lower Middle Terrace (MT 2): 25-30 m	"UMT" at 25-30 m. Well developed at Humanes, Carracosa and the Cañamares-Bornova interfluve at +25-30 m
High Terrace: 40-50 m	Upper Middle Terrace (MT 1): 70-90 m	"OMT" at 60-80 m. Below Jadraque on the right bank at +60-80 m; +60 m at the Jarama confluence
"D" Surface?	Lower High Terrace (HT 2): 110-115 m	"UHT" at 100 m. Recorded at the Cañamares-Bornova interfluve between 880-890 m; +80-100 m at the Jarama confluence
	Upper High Terrace (HT 1): 130-135 m	"OHT" at 110-140 m. At the Cañamares-Bornova interfluve between 913 and 927 m; above Carracosa at 902 to 920 m; at the Jarama confluence at 700 to 720 m (+120-150 m)

In overview, Schwenzner's treatment of the Rio Henares terraces must be viewed with reservation. The omission of sedimentological descriptions and the ambiguous descriptions of individual terrace locations as shown in longitudinal profile suggest that discontinuous terrace segments were interrelated solely on the basis of elevation, a procedure which at best is speculative and in the case of supposedly tectonically produced terraces, would be unacceptable. In addition, the establishment of relationships between the individual terrace remnants recognized by Schwenzner and those noted in this study are frequently compromised or precluded by the vagueness of his locational descriptions.

---

built, a 30 to 40 m Lower Middle Terrace ("UMT") "observed" at Humanes, a 60 to 80 m Upper Middle Terrace ("OMT") on the right side of the valley at Humanes, and a 100 to 110 m Lower High Terrace ("UHT") at 800 to 820 m on the Humanes-Tamajón road. The road referred to in the last case is actually between 760 and 780 m, and the "station" Low Terrace is, in fact, an Upper Middle Terrace. Consequently, it is not clear which features Schwenzner identified.

The lower Henares

Downstream from the Sorbe-Henares confluence near Humanes, broad terrace platforms are entrenched with distinct morphologic definition along both banks of the Rio
Henares. This enables them to be followed subcontinuously downstream to the Henares-
Jarama confluence near Alcalá de Henares. Throughout this lower segment of the river,
the Henares has dissected and broadened its valley into the Tertiary infill of the Tajo
Basin. In contradistinction to the situation along the middle and upper courses, clear
morphologic preservation of these features invites the reconstruction of terrace levels on
the basis of elevation. Such relationships should be considered as tentative, however,
pending sedimentological and/or paleontological corroboration.

At least five terrace levels can be identified at Humanes; these have been mapped
in part by de la Concha (1963). Each terrace is characterized by massive deposits of
subrounded-to-rounded, coarse, quartzite gravels so that definite identifications of particular levels ultimately should be determined by mineralogic and morphometric gravel
analyses. At this time, the following résumé may suffice:

Lower Low Terrace (LT 2, 8 m). This occurs only locally as a platform immediately adjoining the modern channel and within the river bottom land. A prominent remnant of this level occurs at the Torija-Humanes highway bridge over the
Rio Henares.

Upper Low Terrace (LT 1, 15-20 m). This terrace occurs between 710 and 720 m
near the Sorbe-Henares confluence and corresponds to the "T_1" level noted by
de la Concha (1963). Schwenzner (1937) makes no reference to this terrace
although it is presumably shown in Profile 7.

Lower Middle Terrace (MT 2, 20-30 m). Recorded between 720 and 730 m, this
is the "T_2" level partially mapped by de la Concha (1963) that is reached where
Torija-Humanes road and the Canal de Henares intersect. No reference is made
by Schwenzner (1937).

Upper Middle Terrace (MT 1, 40-55 m). This terrace, between 730 and 745 m is
mapped as "T_3" by de la Concha (1963) and referred to by Schwenzner (1937) as a
15 to 20 m Low Terrace on which the Humanes railroad station is built.

Lower High Terrace (HT 2, 70-80 m). This level is marked by gravel deposits
between Km 21 and 22.5 of the Guadalajara-Tamajón road at 760 to 770 m. No
reference is made to this platform either by de la Concha (1963) or by Schwenzner
(1937).

Upper High Terrace (HT 1, 90 m). The Guadalajara-Tamajón road ascends this
terrace between Km 20.1 and 20.0, at 780 m. It is not clear if this corresponds
to Schwenzner's Upper High Terrace.

Preliminary indications, based upon cursory field reconnaissance, are that these
terrace levels can easily be followed and mapped to the Jarama-Henares but at this time
the profiles reconstructed by Schwenzner provide the only basis for interrelating the
Henares-Jarama terraces. Establishment of these relationships, as well as the appropriate correlations of all of the middle and lower Henares terraces, is essential if the

terraces of the Rio Manzanares are to be correlated with the upper Henares and upper Jalón terraces. Riba (1957) has summarized the chronology of the Manzanares terraces with their mammalian fossils and attempted to relate these levels to the Rio Jarama terraces (see also de Terra, 1956). As noted in Table 12, Schwenzner related the three highest terraces of the middle Henares to the three respective high terraces of the Rio Jarama, but in light of the questionable procedures implemented, such correlations must be considered tentative.

The upper Jalón

Fluvial terraces have been noted by several authors but as yet no systematic mapping of Pleistocene deposits has been carried out for the upper Rio Jalón basin. Terrace analogies with the Alto Henares derive from the headwater disposition of these watersheds astride the continental drainage divide formed by the Hesperian Meseta.

Schwenzner (1937:84f) cited a terrace suite preserved along a Rio Jalón left bank arroyo, between Somaen and Arcos de Jalón. Fluvial gravels are noted mantling platforms at +10 to 20 m, +30 to 50 m, +80 to 85 m, and +110 to 120 m. Sánchez de la Torre (1963) identified and mapped +18 to 20 m and +35 to 40 m gravel terraces in the vicinity of Arcos de Jalón that correspond to the two lower levels noted by Schwenzner. Upstream from the gorge cut by the Rio Jalón through the deformed northeastern margin of the Mesozoic uplands, Butzer (1965) has made a detailed study of a +40 to 45 m "terrace" that comprises the Torralba-Ambrona Lower Paleolithic occupation sites. In addition to these citations, the present author has noted conglomerates in the upper Jalón at +12 m near Salinas de Medinaceli and massive travertine platforms 45 m and 100 m above the Rio Jalón, between Fuencaliente and Salinas de Medinaceli. Two generations of marl and tufa deposits have also been isotopically dated in the vicinity of Esteras de Medina (Butzer, unpublished).

As in the case of the upper Henares, the Rio Jalón headwaters contain a range of Pleistocene deposits characteristic of the Hesperian Meseta as a whole. In addition, the implied terrace stratigraphy, pending more detailed study, is not unlike that of the Henares drainage. In the present writer's opinion, at least three distinct fluvial terraces are indicated offset above the campiña valley bottom: a Low Terrace approximately 12 to 15 m, a Middle Terrace at 30 to 45 m, and a High Terrace at 90 to 100 m. A second 110 to 120 m high terrace as cited by Schwenzner (1937) may also be recorded. Isotopic dating of calcrete crust near Contamina and tufa/marl near Esteras de Medina indicates a Würm age for the Low Terrace complex of the middle and upper Jalón respectively (Butzer, unpublished). Interpretation of paleontological, palynological and geomorphological data at the Ambrona-Torralba sites implies a Mindel age for the upper Jalón 30 to 45 m terrace.

Evidence of the last category derives from indications that occupation of the sites ante-
dates two interglacials, the first recorded by a well-developed terra fusca and the second
by later swamp valley-fill beds with a typical interglacial pollen profile (Butzer, 1965;
1969). [6]

Conclusions and Comments

Four terrace levels documented throughout the Alto Henares drainage basin span
most of Pleistocene time. These features comprise deposits which indicate that aggradation occurred during the colder glacial stages. The most recent terrace, referred to as
the Campiña Terrace, comprises the nearly-flat valley bottom adjacent to the present
river courses; radiocarbon dating shows that this surface derives from Holocene time.
The Low Terrace, offset approximately 6 to 12 m above the campiña, can be shown by the
same technique to be of Main Würm age. Immediately downstream from the Mesozoic/
Tertiary geologic contact, valley floors are laterally constricted within gorges so that
Low Terrace deposits are buried under the campiña surface. When this confinement is no
longer present such as in the vicinity of Jadraque, the two levels reassume distinct morphologic expression. Correlations between the terraces of the Alto Henares and the fluvial
platforms recognized by other authors along the middle and lower Henares and along the
upper Jalón of the adjoining drainage basin were attempted. The terrace levels recorded
for the adjacent valleys are shown in Table 13. The results are inconclusive in that the
terrace chronology proposed for the Alto Henares system is in part inconsistent with
these other sequences as presently understood. The relevant issues are summarized in
the following paragraphs.

The proposed correlation of the Alto Henares Middle Terrace deposits, as recorded
at the Riba de Santiuste type-site and elsewhere in the study area, with the Torralba Formation identified and first described by Butzer (1965) is considered to be reliable. This
author has examined the Torralba deposits and Butzer, among others, has studied the
exposure at Riba de Santiuste so that the overall correlation appears confirmed. This conclusion raises other fundamental questions, however, in regard to the Pleistocene history
of the Hesperian Meseta. Foremost of these is the uncertainty that emerges about the
Middle Pleistocene. On the one hand, Low Terrace deposits can be assigned a Würm age
on radiometric grounds, and on the other hand Middle Terrace deposits can be shown to
date from the "Mindel" glacial on account of the correlation indicated with the Torralba

---

[6] Upper Jalón terrace terminology described originally by Butzer (1965) is under
revision for publication in a monograph-length report on the Ambrona-Torralba occupation sites. To avoid confusion the references here use neither the original or revised
terminology but refer to individual terraces only in terms of elevation.

TABLE 13

CORRELATION OF TERRACE SYSTEMS ADJACENT TO THE STUDY AREA*

Manzanares	Lower Jarama		Lower Henares		Alto Henares
Riba (1957)	Riba (1957)	Vaudour (1969)	Schwenzner (1937)	Vaudour (1969)	This Study
		2-3 m (Holocene)			5-7 m (Holocene)
4-8 m 10-12 m (Würm)	8-12 m (Würm)	12 m 18-20 m (Würm)	15-20 m	15 m (Würm)	6-12 m (Würm)
15-25 m (Riss)	15-25 m (Riss)	30 m (Riss)	25-30 m	30 m (Riss)	18-35 m Mid Pleistocene
+45 m (Mindel)	60-80 m (Mindel)	60-70 m (Mindel)	60-80 m	50 m (Mindel)	40-50 m Early Pleistocene
		85 m (Günz)	100 m	80 m; 110 m & 130 m (Günz)	
		+128 m (raña ?)	110-140 m	150-160 m (pre-Günz)	

*Stratigraphic assignments as given by the respective authors.

and Ambrona stratigraphy. Furthermore, recently available Würm radiocarbon dates for the upper Jalón Low Terrace demonstrate a similar Würm-Mindel gap for that basin (see also discussion in Butzer, 1969).[7] Consequently, the disposition of Riss-age legacies in the Hesperian Meseta is enigmatic but this period is not entirely without record, since the disconformity between the typical, shallow Middle Terrace conglomerates and the massive sedimentological sequences at Riba de Santiuste, Riotovi or Horna may represent such a hiatus. It is also possible that the occurrences of bench levels cut into the Middle Terrace may provide a record of this time but, until additional data from other lines of investigation become available, the regional stratigraphy of the Hesperian Meseta must remain incomplete. For these reasons the Middle and High Terrace features of the Alto Henares have been tentatively assigned general mid- and early Pleistocene ages respectively.

It must be emphasized that the suggested correlation of the Alto Henares with the middle and lower Henares terraces is tentative as well. Campiña and Low Terrace

---

[7]Revised terminology recognizes the Ambrona-Torralba complex as a Middle Terrace, rather than a High Terrace as previously suggested (Butzer, 1965).

features of the Alto Henares can be followed subcontinuously downstream to the environs of Jadraque, but firm correlations must await radiocarbon dates from the middle Henares, detailed sedimentological study, and complete gravel analyses. In addition, multiple terrace levels along this segment of the river negate correlations made solely on the basis of altimetric or morphologic relationships and thus underscore the likelihood that the elusive Riss cycle in the Hesperian Meseta is recorded within the Jadraque terrace suite. As noted earlier, in spite of the fact that Schwenzner's study (1937, 50-53) provides a link for relating the Alto Henares terraces to those of the Jadraque vicinity and the Jarama and Manzanares systems, his method of procedure must be questioned.

Unlike the better known Manzanares and Jarama terrace suites, or even the upper Jalón terrace at Torralba and Ambrona, no paleontological evidence has been uncovered in the upper Henares. The single exception occurring in the Middle Terrace in the vicinity of Mojares is uninformative. Discovery of fossil fauna would firmly establish the proposed terrace chronology as well as provide direct and significant links between Paleolithic sites of the Tajo and Ebro Basins.

The freshwater carbonates of the Hesperian Meseta provide a rare opportunity for isotopic dating of a complete terrace suite. As yet, only the Campiña Terrace surface and Low Terrace may be assigned ages with some confidence. Isotopic dating of Middle and High Terrace organic travertines by the Thorium^{230} growth method or perhaps the U^{234} decay method would complete the sequence or at least clearly indicate the relative ages of these terraces. The radiocarbon ages obtained in this study are, nevertheless, the only absolute dates presently available for the Henares Basin.

Pollen analysis has not been completed for the study area. Although the focus of this study is not on paleoecology or Pleistocene environments, pollen spectra and diagrams would throw considerable light on the paleomorphogenetic environments as well as provide valuable correlations with recently acquired pollen data from the adjoining Jalón basin of the Hesperian Meseta (see Butzer, 1964b; Howell, 1966). Availability of these corroborative lines of evidence, in addition to the systematic study of the middle and lower Henares terraces, should contribute significantly to an understanding of the Pleistocene paleoenvironments in central Spain as well as establish direct relationships between the Lower Paleolithic occupation sites of the Jarama/Manzanares system and those of the upper Jalón valley.

APPENDICES

APPENDIX A

LABORATORY PROCEDURES AND TEXTURAL DATA

Sample Procedure

Sediment samples collected subjectively in the field were judged to be representative of the overall target population of the particular strata. More than 200 of these field samples were selected for laboratory study and reduced to workable 50 to 100 grams sizes through the use of a splitting device with 1/2 inch chutes. The portion to be analyzed was oven dried and weighed prior to analysis.

Calcium carbonate normally was dissolved from acid-reacting samples by the addition of 20% hydrochloric acid to a sediment-water slurry. Limestone or dolomite grit or shell material larger than 2 mm was removed from a sample prior to HCl treatment, but this fraction was included when determining total particle size distributions. In this way total HCl soluble content could be determined directly in contrast to the Chittick gasometric procedure which derives calcite or dolomite (or carbonate) content by tabular transformations based upon the volume of CO_2 released after the measured addition of sulfuric or dilute hydrochloric acids. Although the latter method proves useful for samples with comparatively low calcium carbonate contents, the procedure is appreciably more time-consuming and less accurate for samples with high calcium carbonate contents. The direct addition of dilute HCl proved to be of further value in those cases where the non-soluble residual of biogenic deposits was of interest. After the complete removal of HCl soluble materials, the individual samples were thoroughly washed, oven dried and reweighed to determine percentage of mass loss by weight.

Clastic samples were more difficult to analyze. These conglomerate and breccia materials in most cases were cemented by calcite or indurated, such that specimens could not be broken down for particle size analysis without chemical treatment. Several procedures were attempted involving solutions of hydrochloric and sulfuric acids in varying strengths, in which samples soaked for different time intervals, punctuated by aperiodic oven drying and firm hammer blows. As might be expected, however, those procedures which were in any way effective in weakening the cementation also altered the morphometry of the clastic limestone or dolomite components so that particle size and/or morphometric gravel analyses would be meaningless. Description of these samples is based, therefore, upon unaltered pieces; color notations in the text refer to the predominant color of the matrix and morphometric properties of the larger clastic fragments were determined while intact.

pH Determination

Determinations were made electrometrically with the use of a Beckman pH meter. A paste was prepared using 10 to 20 grams of the field sample and distilled water. Equilibrium intervals between preparation and determination normally involved 15 to 20 minutes.

Texture Analysis

Particle size fractions by weight were determined by the hydrometer method, for particles smaller than 63 microns, and by wet-sieving for the coarser fractions. The standard hydrometer procedure, which determines amount of material remaining in liquid (water) suspension at predetermined lapsed time intervals, is described by the American Society for Testing Materials (A. S. T. M., 1958:1119-29; also A. S. T. M. Supplement Book, 1960:1151f). The amounts of coarser size fractions were determined after completion of the hydrometer process by passing the sample through a nested set of U. S. Standard sieves. The separated fractions subsequently were decanted, oven dried and weighed.

The particle size classes used throughout this study for a particular sediment are defined as follows:

(1) granule	more than 2 mm
(2) coarse sand	595–2000 microns
(3) medium coarse sand	210–595 microns
(4) medium sand	63–210 microns
(5) fine sand	20–63 microns
(6) coarse silt	6–20 microns
(7) fine silt	2–6 microns
(8) clay	less than 2 microns

This is a non-logarithmic scheme modified from the Atterberg scale. The percentage content (by weight) of non-carbonate sediment in each of these classes is shown in Table 14 for samples cited in the text. In many cases, particularly marls, amounts of non-carbonate residual material were insufficient to permit meaningful particle size analysis, since the hydrometer method generally is impractical and unreliable for concentrations of approximately 6 gr/liter or less. In these instances, texture was determined subjectively by careful examination and "feel" of the sediment.

Text Descriptions

Notations of sediment or soil color are according to the Munsell Soil Color Chart (1954). The degree of sorting refers to the concentration of the silt and sand particles (2-2000 microns) into size classes. As defined by Payne (1942), sorting is "good" when one or two size grades comprise 90% of the sample; "moderate, " if three or four size grades comprise 90%; and "poor" if five or six size grades are required to include 90% of the sample. Terminology used to describe soil/sediment structure and configuration of horizon/strata contact boundaries follows that of the Seventh Approximation of the U.S. Soil Survey (1960). Stratification refers to the degree of definition of bedding planes in a particular strata; the terminology used here is solely comparative, varying from unstratified to well stratified depending upon the conspicuousness of the bedding.

The coherence or hardness of the soil or sediment is defined in terms of consolidation in a dry state. Three degrees are recognized in the text: unconsolidated (in the sense of loose or amorphous), weakly consolidated (semi- or loosely consolidated, or indurated) which yields with difficulty under hand pressure, and well consolidated (or indurated) which breaks only with a hammer. Cementation terminology is invoked when the hardness property of the sediment derives from an agent of adhesion, usually calcium carbonate or in some cases silica. The terms used are relative: weakly or semi-cemented materials can be broken in hand, but with difficulty; cemented sediments yield only to a hammer; and well-cemented materials yield only with repeated and firm hammer blows.

Textural class description of mechanically analyzed samples is based upon the relative amounts of the sand (including granule grade), silt and clay components. The definition of grade-size categories generally follows the system used by Wentworth (1922):

Grade size	Percentage Content	Texture-class
Sand	$>$ Gravel ($>$10), and others $>$10	Gravelly sand
Sand	$>$80	Sand
Sand	$>$Silt ($>$10), and others $>$10	Silty sand
Silt	$>$Sand ($>$10), and others $>$10	Sandy silt
Silt	$>$80	Silt
Silt	$>$Clay ($>$10), and others $>$10	Clayey silt
Clay	$>$Silt ($>$10), and others $>$10	Silty clay
Clay	$>$80	Clay

In the case of a marl, the sediment may contain a high percentage of fine sand in addition to the silt and/or clay, but must also have a high carbonate content (always greater than 30% by weight, but usually greater than 50%). When such a sediment contains appreciable amounts (usually more than 20%) of medium sand or coarser sand (i.e., $>$63 microns), it may be referred to as a sandy marl.

TABLE 14

TEXTURAL DATA OF NON-CARBONATE RESIDUE OF SAMPLES PRESENTED IN THE TEXT TABLES

Table 4

Sample no.	(1)	(2)	(3)	(4)	(5)	(6)	(7)	(8)
H615	0.1	1.7	8.3	17.5	28.2	15.6	14.7	23.9
1425			(Insufficient residue--0.1 gr--for particle size analysis)					
1426			(Insufficient residue--5.9 gr--for particle size analysis)					
1427			(Insufficient residue--0.4 gr--for particle size analysis)					
1428			(Insufficient residue--6.0 gr--for particle size analysis)					
1430			(Insufficient residue--0.9 gr--for particle size analysis)					
1431a	–	0.1	1.9	10.2	48.0	11.4	10.2	18.2
1431b	–	0.8	2.8	25.6	58.0	3.2	6.4	3.2
1431c			(Insufficient residue--5.4 gr--for particle size analysis)					
1432	–	–	0.2	3.6	44.6	18.3	11.2	22.1
1436	0.4	2.5	3.5	14.2	19.0	8.4	15.8	36.2
1438			(Insufficient residue--6.4 gr--for particle size analysis)					

Table 5

(Cemented clastic deposits; no particle size analysis)

Table 6

Sample no.	(1)	(2)	(3)	(4)	(5)	(6)	(7)	(8)
H211	3.2	2.1	1.8	2.6	4.5	1.9	10.9	73.0
H210	–	8.8	31.1	20.6	12.5	5.5	7.5	14.0
H209	–	2.5	9.3	19.7	34.5	14.5	7.5	12.0
H208	–	0.6	0.4	15.3	22.2	8.7	8.1	44.7
H316	33.9	9.5	21.5	12.5	5.3	1.8	3.6	11.9
H315	7.0	28.2	20.3	9.0	4.6	4.6	3.1	23.2
H314	–	1.0	2.5	7.9	24.8	7.3	5.0	51.5
H306	52.0	10.4	18.4	5.2	5.0	2.5	2.5	4.0
H305	–	6.0	9.8	10.7	29.0	15.0	10.5	19.0

H313	40.7	13.5	17.8	24.5	3.5	-	-	-
H312	23.0	3.8	3.8	10.5	31.8	27.1	-	-
H311	61.9	9.1	10.9	18.1	-	-	-	-
H310	48.0	12.5	12.5	24.1	2.9	-	-	-
H309	64.0	12.2	8.1	14.9	0.8	-	-	-
H308	65.0	5.0	10.0	19.9	0.1	-	-	-
H302	(Insufficient residue—0.8 gr—for particle size analysis)							

Table 7

S120	3.3	1.0	1.0	2.0	2.8	5.6	3.8	80.5
S122	32.8	9.5	10.6	30.5	13.2	3.4	-	-
S123	47.5	10.0	8.5	21.0	8.0	5.0	-	-
S124	27.8	8.7	11.8	33.2	16.5	2.0	-	-
S126	8.6	-	2.8	2.8	4.9	11.8	7.2	61.9
S99	34.5	9.6	12.8	31.2	8.4	3.5	-	-
S100A	8.5	5.3	5.3	7.7	8.3	3.8	1.0	60.1
S98	12.8	-	1.8	-	1.5	3.1	2.6	78.2
S97	26.6	7.8	7.8	30.7	25.6	1.5	-	-
S104	7.0	1.0	1.5	3.0	3.5	5.5	5.3	73.2
S105	15.5	11.0	11.0	48.0	8.1	2.8	3.6	-
S107	5.5	-	2.5	3.0	4.7	7.0	5.6	68.7
S109	12.0	-	3.5	7.0	7.5	8.0	2.6	59.4
S110	17.5	8.5	12.5	40.0	19.6	1.9	-	-
S111	28.0	12.0	16.0	28.0	7.1	1.0	1.9	6.0
S96	31.0	7.3	11.3	28.9	20.0	1.5	-	-
S113	10.0	1.0	1.0	3.0	3.0	6.2	7.6	68.2
S115	21.5	10.5	12.5	25.0	24.8	5.7	-	-
S204	3.6	1.2	1.2	5.3	1.0	1.0	1.0	85.7
S215	30.0	8.5	5.6	37.5	15.4	3.0	-	-
S203	22.5	14.5	11.5	40.5	10.0	1.0	-	-
S213	17.1	5.7	10.6	31.4	24.2	9.5	1.5	-
S212	43.0	6.0	17.0	21.6	8.5	3.4	0.5	-
S216	31.0	18.6	12.4	31.0	5.0	1.0	1.0	-
S208	3.0	2.0	-	4.0	1.0	0.8	1.3	87.9

TABLE 14--Continued

Sample no.	(1)	(2)	(3)	(4)	(5)	(6)	(7)	(8)
S311	1.0	2.4	3.9	16.3	32.8	5.8	8.9	28.9
S310	53.8	3.8	9.6	9.8	9.0	5.5	2.5	6.0

Table 8

Sample no.	(1)	(2)	(3)	(4)	(5)	(6)	(7)	(8)
D703	(Insufficient residue--0.4 gr--for particle size analysis)							
D702	(Insufficient residue--0.2 gr--for particle size analysis)							
D700A	-	-	7.9	13.7	15.3	15.3	15.3	32.5
D901	-	1.3	7.9	12.5	11.5	10.3	10.3	46.2
D902	50.1	15.6	10.8	7.2	2.3	-	1.2	12.8
D904	-	1.0	4.9	7.5	16.7	8.9	9.8	51.2
D905	-	1.0	5.3	13.6	18.0	11.7	8.2	42.2
D906	71.4	1.0	4.1	4.1	1.7	1.7	-	16.0
D907	-	2.0	12.0	19.0	17.5	13.5	7.8	28.2
D903	-	2.2	8.8	20.3	15.7	7.9	5.2	39.9
D303	75.1	2.5	5.5	4.9	4.0	1.5	1.0	5.5
D204	73.4	2.9	9.5	7.2	1.5	1.0	1.0	3.5
D203	17.6	4.1	18.7	21.6	11.5	9.0	6.0	11.5
D301	82.0	2.4	6.5	3.6	1.5	2.0	-	2.0
D201	30.8	3.9	11.8	14.0	8.5	10.0	5.0	16.0
D501	72.6	3.2	8.2	6.0	1.0	2.5	1.5	5.0
D502	-	6.2	5.4	11.4	35.5	15.5	13.5	13.5
D503	-	-	-	0.5	11.5	20.5	10.5	57.0
D504	-	-	-	1.5	14.0	7.0	14.0	63.5
D505	-	1.6	8.9	19.0	28.5	14.5	11.0	16.5

APPENDIX B

GRAVEL ANALYSES

Gravel shape description is based upon a modified version of the Lüttig method outlined by Butzer (1964, 160-64). The measures include statements of pebble rounding, shape and size. The index of rounding (ρ) is the estimated percentage of convex surface of the major circumference of the pebble. Mean rounding values for individual samples can be given verbal expression according to the following classes:

0-10%	angular
11-20%	subangular
21-40%	subrounded
41-60%	rounded
>60%	well-rounded

Pebbles with a rounding index of 8% or less comprise the detrital component. The degree of homogeneity or scatter of rounding values within a sample is expressed by the coefficient of variation (CV), which is derived by dividing the standard deviation of the sample by the sample mean, multiplied by 100. The result allows the gravel sample to be classed as:

very homogeneous	0-25
homogeneous	25-50
heterogeneous	50-75
very heterogeneous	>75

Pebble shape is defined by ratios of the major or minor axes and the pebble thickness. Both the thickness/major axis (E/L) and thickness/minor axis (E/l) ratios are given as mean values for each sample. Higher values of these ratios, generally above 50% and 65% respectively, reflect a predominant rounding tendency rather than flattening, which is presumed to be an indication of a preponderance of sliding transport motion.

Pebble size, given in centimeters, refers to the average length of the major axis, while the composition of each sample is given as the percentage of quartzite, limestone (or dolomite), and other (sandstone) lithologies, in that order.

Gravel morphometry is used here only for descriptive purposes; no transport inferences are offered or warranted. The quartzite pebbles are derived from the Buntsandstein or Tertiary conglomerates that are rounded or well-rounded and coarse, so that the morphometry of Pleistocene gravels is largely inherited. The quartzite characteristically is brittle and fractures easily upon mechanical impact so that implications of

189

TABLE 15

GRAVEL MORPHOMETRY

Location	Roundness			Shape		Size		Sample		
	Mean ρ	CV of ρ	Percentage of Detritus	E/L	E/l	Mean Length in cm	Size	(1)	Composition[a] (2)	(3)
Rio Henares										
Low Terrace at Arroyo de la Calera	48.6	37.3	–	54.2	78.2	3.8	49	96	4	–
Low Terrace at Arroyo de la Calera	37.7	47.5	–	51.4	79.0	4.9	35	100	–	–
Low Terrace at Baides	51.8	32.6	–	47.3	77.0	4.0	32	27	–73	–
Middle Terrace at Arroyo de la Calera	43.1	45.3	–	59.1	81.2	2.5	52	27	73	–
Middle Terrace at Arroyo de la Calera	43.4	39.1	–	59.3	80.4	3.5	50	100	–	–
Middle Terrace at Sigüenza	54.3	32.8	–	62.1	78.5	3.1	32	100	–	–
High Terrace at Arroyo de la Calera	30.3	49.3	3.0	57.7	76.1	4.2	36	100	–	–
Rio Salado										
Low Terrace near Salinas de la Olmedilla	30.2	31.2	5.0	50.6	73.6	1.7	59	–	100	–
Low Terrace near Imón	43.9	44.7	1.5	50.2	78.0	1.8	62	100	–	–

Low Terrace near Viana de Jadraque	41.0	43.4	1.5	46.3	77.1	3.5	59	91	9	–
Middle Terrace at Riba de Santiuste: Horizontal sequence, Bed a	31.3	13.3	–	37.7	68.3	4.3	35	74	26	–
Middle Terrace at Riba de Santiuste: Horizontal sequence, Bed e	31.4	39.9	–	40.0	70.8	4.2	31	55	42	3
Middle Terrace at Riba de Santiuste: Inclined sequence, Bed e	29.5	31.0	–	37.3	65.0	4.2	33	64	36	–
Middle Terrace at Riba de Santiuste: Inclined sequence, Bed g	33.0	43.7	–	35.8	72.0	3.7	29	100	–	–
Middle Terrace at Riba de Santiuste: Inclined sequence, Bed h	27.9	21.3	–	46.4	86.9	3.9	35	72	28	–
Middle Terrace at Riotovi: Bed a	25.6	17.9	–	50.2	75.7	3.4	54	100	–	–
Middle Terrace at Riotovi: Bed g	30.6	32.3	–	56.4	83.0	3.0	29	5	95	–
Slope Deposits at Paredes de Sigüenza	26.5	66.0	9.0	49.5	80.4	2.8	57	18	82	–

192

TABLE 14--Continued

Location	Roundness			Shape		Size		Sample			
	Mean ρ	CV of ρ	Percentage of Detritus	E/L	E/l	Mean Length in cm	Size	Composition[a]			
								(1)	(2)	(3)	

Rio Dulce

Location	Mean ρ	CV of ρ	Percentage of Detritus	E/L	E/l	Mean Length in cm	Size	(1)	(2)	(3)
Campiña Terrace at Mandayona	60.7	26.9	-	50.3	77.3	3.3	54	100	-	-
Low Terrace at Mandayona	37.7	91.3	-	53.4	79.6	3.0	50	1	99	-
Middle Terrace at Mandayona	19.4	22.0	39.0	44.5	72.5	4.3	31	6	94	-

[a] Percentage of (1) quartzite, (2) limestone or dolomite, and (3) other lithologies that comprise the sample.

frost-shattering are uncertain. In some cases, quartzite fracturing is sufficiently prominent to effect lower mean rounding values. Limestone pebbles frequently show evidence
of solution and consequently the relative contributions of weathering and transport motion
to pebble morphology are often difficult to assess.

APPENDIX C

RADIOCARBON DATES

More than a dozen samples were submitted to Isotopes Inc. for radiocarbon dates. Eleven of these were suitable for processing, eight of which were collected throughout the Alto Henares drainage area and are inventoried below. Three additional samples from the Rio Jalón basin will be reported on separately by K. W. Butzer. Care was taken during sample collection and storage to insure against contamination. All samples were cleaned with dilute hydrochloric acid in the lab to remove extraneous matter; the single organic sample was not pretreated with sodium hydroxide since tests showed that nearly all of the organic carbon was removed by such treatment. The relatively small standard deviations of each sample indicate that mechanical errors are minimal and sufficient quantities of sample material could be processed.

Ages of the samples tend to cluster about several time ranges: only one sample is in the order of 30,000 years old, but two are approximately 25,000 years old, three in the 20,000 year range, three in the 10,000 year range, and the two youngest are about 5000 years. The fact that these samples were collected from stratigraphic contexts scattered through three basins speaks for their general reliability, and comparative dates (see I-4593 and 4594 below) from carbonaceous soil and calcium carbonate suggest strongly that the inorganic carbonates dated had ample opportunity for equilibrium with atmospheric carbon prior to their precipitation by stream or ponded waters. It is felt that "dead" carbonates from the limestone/dolomite bedrock are of limited significance.

Isotopes, Inc. Sample Number	Sample Description	Sample Age (in years)
I-3110	From intact, porous, organic tufa beds at the top of the Rio Henares Low Terrace type-site upstream from Baides. The sample age dates a waning accretionary phase of the beds that rest disconformably upon older deposits below. The Holocene date records the local alluviation contemporary with the campiña surface development of the upper Henares proper, upstream from the Mesozoic-Tertiary geologic contact.	6560±130
I-3111	From the uppermost meter of more than 8 m of massive organic tufa below the village of Horna. These deposits are the youngest facies of a lacustrine-type sequence recorded at this site and are considered to be contemporary with the development of the upper Rio Henares Low Terrace. Elsewhere (Arroyo de la Calera) this time period is recorded by soil development (charcoal collected insufficient for date).	30,650±1350
I-3112	Porous tufa obtained from the upper meter of beds exposed above the Rio Dulce right bank at the Mandayona-Sigüenza road bridge. Outcrops of the same material reoccur all along this segment of the river, such as at the terrace type-site described 1 km farther downstream. The Holocene date is considered to record alluviation of the Rio Dulce campiña surface and, together with Sample I-3110, provides a firm stratigraphic marker both in the upper Henares and Dulce basins.	4020±110
I-3543	From a cemented, very pale brown, sandy marl found intercalated within 2 m of basal gravels of the campiña surface type-site near Mandayona. The lens is approximately 1.50 m above mean river level. There are no disconformities within the gravel bed or above it so that the date is considered to represent an interval of ponding, with restricted bed-load transport.	19,700±400
I-3544	From an unconsolidated white marl (99.5% carbonates) in the upper Rio Dulce basin. This bed occurs in proximity to an older, massive-to-laminated travertine platform. The dated material does not occur within a local terrace context but nonetheless is similar to and contemporary with deposits in the nearby upper Jalón valley.	25,300±750
I-3545	From a thin calcareous sandy band within the Henares Low Terrace type-site, upstream from Baides. It is separated from the higher and younger (I-3110) tufa beds that complete this sequence by two disconformities. The comparable ages of this material and the Rio Dulce campiña deposits (I-3543) establish the interrelationship of these two features (Table 11).	19,450±350
I-4593	Carbonate fraction from a buried organic horizon within the Rio Dulce campiña sequence upstream from Mandayona. The organic muck contained small charcoal fragments, gastropod shells, and inorganic carbonates. These component materials were sufficiently abundant to	12,570±190

Isotopes, Inc. Sample Number	Sample Description	Sample Age (in years)
	permit separation of the carbonate and organic constituents and separate radiocarbon dates.	
I-4594	Carbonaceous soil from the preceding sample (I-4593, carbonate). The age is ±20% younger, but since finely divided organic carbon is highly susceptible to contamination by younger humus, the carbonate date probably provides a closer approximation of true age (ca. 12,000 B.P.).	9750±160

LIST OF REFERENCES

LIST OF REFERENCES

ALVIRA, T., and GUERRA, A. (1951) Estudio de algunos suelos de las cuencas del Tajo y del Duero. Anal. Edaf. Agrob., v. 10:51-65.

AMERICAN SOCIETY FOR TESTING MATERIALS. (1958) Procedures, Part 4:1119-29.

ARAMBOURG, C. (1952) The Red Beds of the Mediterranean Basin. Proc. I Pan-Af. Conf. Prehist., Nairobi, 1947:39-45.

ASENSIO AMOR, I. (1966) El sistema morfogenético fluvio-torrencial en la zona meridional de la Sierra de Gredos. Estud. Geogr., v. 102:53-74.

BAILEY, R. G., and RICE, R. M. (1969) Soil slippage: an indicator of slope stability on chaparral watersheds of southern California. Prof. Geog., v. 21(3):172-77.

BENNETT, H. H. (1960) Soil Erosion in Spain. Geog Rev., v. 50:59-72.

BIROT, P. (1933) Le Relief de la Sierra de Alto Rey et sa bordure orientale. Bull. Assoc. Géogr. Français, No. 70:92-98.

_____. (1945) Sobre la morfología del segmento occidental de la Sierra del Guadarrama. Estud. Geogr., v. 18:155-68.

BIROT, P., and SOLÉ SABARIS, L. (1951) Sobre un rasgo morfológico paradójico de los macizos cristalinos de la Cordillera Central Ibérica. Estud. Geogr., v. 45: 807-13.

BIROT, P., and SOLÉ SABARIS, L. (1954) Investigaciones sobre morfología de la Cordillera Central española. C.S.I.C. Inst. Seb. Elcano, Madrid.

BOLETÍN MENSUAL CLIMATOLÓGICO. (1957-1966.) Madrid, Servicio Meteorológico.

BONATTI, E. (1966) Northern Mediterranean climate during the last Würm glaciation. Nature, v. 209:984-85.

BONIFAY, E. (1957) Age et signification des sols rouges méditerranéens en Provence. C.R. Acad. Sci., v. 244:3075-77.

_____. (1959) Stratigraphie des loess Würmiens en Provence. C.R. Acad. Sci., v. 248:123-25.

_____. (1962) Quaternaire et préhistoire des régions méditerranéenes Françaises. Quaternaria VI:343-70.

BRINKMAN, R. (1960) Geologic evolution of Europe. Stuttgart, Ferdinand Enke.

BRITISH METEOROLOGICAL OFFICE. (1958) Tables of temperature, relative humidity and precipitation for the World--Part IV, London.

BRUNNACKER, K., and LOŽEK, V. (1969) Löss-vorkommen in südostspanien. Zeit. f. Geomorph., v. 13(3):297-316.

BUTZER, K. W. (1961) Paleoclimatic implications of Pleistocene stratigraphy in the Mediterranean Area. Ann. N.Y. Acad. Sci., v. 95(1):449-56.

_____. (1963a) Climatic-geomorphologic interpretation of Pleistocene sediments in the Eurafrican subtropics in African ecology and human evolution. Howell, F. C., and Bourliere, F. (eds.), Chicago, Aldine.

_____. (1963b) The last "pluvial" phase in the Eurafrican subtropics. Arid Zone Res. (UNESCO), v. 20:211-21.

_____. (1964a) Pleistocene geomorphology and stratigraphy of the Costa Brava region (Catalonia). Abhl. Akad. Wiss. Lit. (Mainz), Math.-Naturw. Kl., No. 1.

_____. (1964b) Environment and archeology. Chicago, Aldine.

_____. (1965) Acheulian occupation sites at Torralba and Ambrona, Spain: their geology. Science, v. 150(3704):1718-22.

_____. (1969) Comments on: Culture traditions and environment of early man, by D. Collins. Current Anthropology, v. 10:303-416.

BUTZER, K. W., and CUERDA, J. (1962) Coastal stratigraphy of southern Mallorca and its implications for the Pleistocene chronology of the Mediterranean Sea. J. of Geol., v. 20:398-416.

BUTZER, K. W., and HANSEN, C. L. (1968) Desert and river in Nubia. Madison, Univ. of Wisc. Press.

CASTELL, J., and de la CONCHA, Serafín. (1956a) Explicación hoja número 434, Barahona. Inst. Geol. y Min. España.

_____. (1956b) Explicación hoja número 462, Maranchón. Inst. Geol. y Min. España.

_____. (1959) Explicación hoja número 435, Arcos de Jalón. Inst. Geol. y Min. España.

CEBALLOS, Luis (ed.). (1966) Mapa Forestal de España, 1:400,000. Ministerio de Agricultura, Madrid.

de la CONCHA, Serafín. (1963) Explicación hoja número 486, Jadraque. Inst. Geol. y Min. España.

CRUSAFONT PAIRÓ, M.; MELÉNDEZ, B.; and TRUYOLS SANTONJA, J. (1960) El yacimiento de vertebrados de Huérmeces del Cerro (Guadalajara) y su significado cronoestratigráfico. Estud. Geol., v. 16:243-54.

CRUSAFONT PAIRÓ, M., and TRUYOLS SANTONJA, J. (1960) El Mioceno de las cuencas de Castilla y de la Cordillera Ibérica. Notas y Com. del Inst. Geol. y Min. de España, No. 60:127-40.

DANTÍN, J. (1924) Acerca de un molar de Listriedon splendens V. Meyer hallado en Jadraque (Guadalajara). As. Esp. Prog. Cienc., Congreso de Salamanca.

DURAND, J. H. (1959) Les sols rouges et les croûtes en Algérie. Serv. Etudes Sci., Alger-Birmandreis.

EMBLETON, C., and KING, C. A. M. (1968) Glacial and periglacial geomorphology. New York, St. Martin's.

ERN, H. (1966) Die dreidimensionale Anordung der Gebirgsvegetation auf der Iberischen Halbinsel. Bonner Geogr. Abhl., v. 37.

EYRE, S. R. (1963) Vegetation and soils. Chicago, Aldine.

FRÄNZLE, O. (1959) Glaziale und periglaziale Formbildung im östlichen Kastilischen Scheidegebirge. Bonner Geogr. Abhl., v. 26.

GUERRA, A. (1968) Mapa de suelos de España, escala 1:1,000,000 y descripción de las asocianes y tipos principales de suelos. Inst. Nac. de Edaf. y Agrob.

HAMMOND, E. H. (1964) Analysis of properties in Land Form Geography: an application to broad-scale land form mapping. Ann. Assoc. Am. Geog., v. 54:11-19.

HERNÁNDEZ PACHECO, E. (1934) Síntesis fisiográfica y geológica de España. Trab. Mus. Cienc. Nat., Ser. Geol. No. 38.

HERNÁNDEZ PACHECO, F. (1932) Tres ciclos de erosión geológica en las sierras orientales de la Cordillera Central. Bol. R. Soc. Esp. Hist. Nat., v. 32:455-60.

_____. (1962) La formación o depositos de grandes bloques de edad Plioceno, su relación con la Raña. Estud. Geol., v. 18:75-88.

HESSINGER, E. (1949) La distribución estacional de las Precipitaciones en la Península Ibérica y sus causas. Estud. Geogr., v. 10:59-128.

HOPFNER, H. (1954) La Evolución de los bosques de Castilla la Vieja en tiempos históricos. Estud. Geogr., v. 15:415-30.

HOUSTON, J. M. (1964) The western Mediterranean world. London, Longmans.

HOWELL, F. C. (1966) Observations on the earlier phases of the European Lower Paleolithic in Recent studies in Paleoanthropology, Clark, J. D., and Howell, F. C. (eds.). Am. Anth. Assoc., v. 68:88-201.

JORDANA, L. (1935) Breve reseña físico-geológica de la provincia de Guadalajara. Inst. Geol. y Min. España.

JORDANA, L., and KINDELÁN, J. A. (1951) Explicación hoja número 461, Sigüenza. Inst. Geol. y Min. España.

KELLER, W. D. (1957) The principles of chemical weathering. Columbia, Mo., Lucas Brothers.

KLINGE, H. (1958) Eine Stellungrahme zur Altersfrage von Terra-Rossa-Vorkommen. Z. Pfl. Ernahr., Düng., Bodenk., v. 81:56-63.

KUBIENA, W. L. (1953) The soils of Europe. London, Murby.

KUBIENA, W. L. (1954) Über Reliktböden in Spanien. Aichinger Festschrift, 1. Mitt. Inst. f. angewandte Vegetationskunde, Vienna:213-24.

_____. (1956) Kurze Übersicht über die wichtigsten Formen der Bodenbildung in Spanien in Die Pflanzen welt Spaniens I, Veröff. Geobot. Inst. Rübel Zürich, v. 31:23-31.

LANDSBERG, H. E.; LIPPMAN, H.; PAFFEN, Kh.; and TROLL, C. (1965) World Maps of Climatology, 2nd. New York, Springer-Verlag.

LAUTENSACH, H. (1960) Die Temperaturverhältnisse der Iberischen Halbinsel und ihr Jahresgang. Die Erde, v. 91:86-114.

_____. (1964) Die Iberische Halbinsel. Munich, Keyser.

LAUTENSACH, H., and MAYER, E. (1961) Iberische Meseta und Iberische Masse. Zeit. f. Geomorph., v. 5(3):161-80.

LEOPOLD, L. B.; EMMETT, W. W.; and MYRICK, R. M. (1966) Channel and hillslope processes in a semiarid area, New Mexico. U.S. Geol. Sur. Prof. Pap. 352-G: 193-253.

LLOPIS LLADÓ, N., and SÁNCHEZ de la TORRE, L. (1965) Mapa geológico de la región de Santamera, 1:25,000. Inst. Geol. Económica.

LÓPEZ GÓMEZ, J. (1959) El clima de España según la clasificación de Köppen. Estud. Geogr., v. 20:167-88.

LOTZE, F. (1929) Stratigraphie und Tektonik des Keltiberischen Grundegebirges. Abhl. d. Ges. der Wiss. Göttingen, v. 14(3).

MAXWELL, J. C. (1967) Quantitative geomorphology of some mountain chapparal watersheds in southern California in Quantitative geography, Part II, Dept. of Geog., Northwestern Univ., Evanston:108-226.

MENSCHING, H. (1956) Karst y terra rossa en Mallorca. Estud. Geogr., v. 65:659-72.

_____. (1967) Die Regionale und Klimatisch-morphologische Differenzierung von Bergflussflächen auf der Iberischen Halbinsel. Würzburger Geographische Arbeiten, v. 12.

MOYA, M., and KINDELÁN, J. A. (1951) Explicación hoja número 487, Ledanca. Inst. Geol. y Min. España.

MUNSELL COLOR CO. (1954) Munsell soil color charts, Baltimore.

de NOVO, P., and CHICARRO, F. (1957) Diccionario de geología y ciencias afines. Editorial Labor, S.A., Madrid.

OLLIER, D. C. (1967) Landform description without stage names. Aust. Geogr. Stud., v. 5:73-80.

PALMER, N. (1965) The occurrence of ground water in limestone. Compass, v. 42(4): 246-55.

PAYNE, T. G. (1942) Stratigraphical analysis and environmental reconstruction. Bull. Am. Assoc. Petroleum Geologists, v. 26:1697-1770.

RIBA, O. (1957) Terrasses de Manzanares et du Jarama aux environs de Madrid. Livret Guide de l'excursion C_2. V Intern. INQUA Congress, Madrid, 1957.

RICHTER, G., and TEICHMÜLLER, R. (1933) Die Entwicklung der Keltiberischen Ketten. Abhl. d. Ges. der Wiss. Göttingen, Kl. III, No. 7.

ROYO GÓMEZ, J. (1928) El Terciario continental de la cuenca alta del Tajo. Explicación hoja 560, Alcalá de Henares. Inst. Geol. y Min. España.

RUTTE, E. (1958) Kalkkrusten in Spanien. Abhl. Neues Jb. Geol. Paläontol., v. 106: 52-138.

SÁNCHEZ de la TORRE, L. (1963) El Borde Mioceno en Arcos de Jalón. Estud. Geogr., v. 19:109-36.

SCHAEFER, I. (1950) Die dilviale Erosion und Akkumulation. Forsch. z. deut. Landes-kunde, 49.

SCHRÖDER, E. (1930) Das Grenzgebiet von Guadarrama und Hesperischen Ketten (Zentralspanien). Abhl. d. Ges. der Wiss. Göttingen, Math.-Phys. Kl., N.F. Bd. 16(3):121-80.

SCHWENZNER, J. E. (1937) Zur Morphologie des Zentralspanischen Hochlandes. Geogr. Abhl., Stuttgart, v. 10.

SERVICIO METEOROLÓGICO NACIONAL. Boletín Mensual Climatológico, Madrid.

SLACK, K. V. (1967) Physical and chemical description of Birch Creek: a travertine depositing stream, Inyo County, Calif. U.S. Geol. Sur. Prof. Pap. 549-A.

SOLÉ SABARIS, L. (1952) España--Geografía Física in Geografía por España y Portugal I, Manuel de Terán (ed.), Barcelona.

_____. (1962) Le quaternaire marin des Baléares et ses rapports avec les côtes méditerranéenes de la Péninsule Ibérique. Quaternaria VI:309-42.

SOS BAYNAT, V. (1957) Observaciones sobre la formación y la edad de los rañas. Cur. y Conf. Inst. Lucas Mallada, v. 4:33-36.

de TERRA, H. (1956) Climatic terraces and the Paleolithic of Spain. Diputación Provincial de Asturias, Servicio de Investigaciones Arqueológicas, Oviedo.

THORNES, J. B. (1967) Some factors affecting erosion and deposition in the Alto Duero, Spain. Unpub. Ph.D. dissertation, Dept. of Geogr., London School of Econ. and Political Science, Univ. of London.

THORNTHWAITE, C. W. (1958) An approach toward a rational classification of climate. Geog Rev., v. 38:55-94.

TRICART, J. (1966) Quelques aspects des phénomènes périglaciaires quaternaires dans la péninsule ibérique. Biul. Peryglac., v. 15:313-27.

204

U. S. Dept. Agriculture. (1960) Soil classification, a comprehensive system, 7th approx. U. S. Soil Survey.

VAUDOUR, J. (1969) Données nouvelles et hypothèses sur le quaternaire de la région de Madrid. Rev. Géogr. des Pays Méd.: 79-92.

del VILLAR, E. H. (1937) Los suelos de la península Luso-Ibérica. Robinson, G. W. (trans.). London, Murby.

WENTWORTH, C. K. (1922) A scale of grade and class terms for clastic sediments. J. of Geol., v. 30:277-92.

WICHE, K. (1961) Beitrage zur Formenentwicklung der Sierren am unteren Segura (Südostspanien). Mitt. Osterr. Geogr. Ges., v. 103:125-57.

_____. (1964) Formen der Pleistozänen Erosion und Akkumulation in Südostspanien. Sixth Int. Cong. on Quat., Warsaw 1961, v. 4:187-97.

ZARANZA, I. C. (1964) Mapa geológico de la región de Alcuneza, 1:25,000. Inst. Geol. Económica.

THE UNIVERSITY OF CHICAGO
DEPARTMENT OF GEOGRAPHY
RESEARCH PAPERS (Lithographed, 6×9 Inches)

(Available from Department of Geography, Rosenwald Hall, The University of Chicago, Chicago Illinois 60637. Price: $4.50 each; by series subscription, $4.00 each.)

*1. GROSS, HERBERT HENRY. *Educational Land Use in the River Forest–Oak Park Community (Illinois)*
*2. EISEN, EDNA E. *Educational Land Use in Lake County, Ohio*
*3. WEIGEND, GUIDO GUSTAV. *The Cultural Pattern of South Tyrol (Italy)*
*4. NELSON, HOWARD JOSEPH. *The Livelihood Structure of Des Moines, Iowa*
*5. MATTHEWS, JAMES SWINTON. *Expressions of Urbanism in the Sequent Occupance of Northeastern Ohio*
*6. GINSBURG, NORTON SYDNEY. *Japanese Prewar Trade and Shipping in the Oriental Triangle*
*7. KEMLER, JOHN H. *The Struggle for Wolfram in the Iberian Peninsula, June, 1942—June, 1944: A Study in Political and Economic Geography in Wartime*
*8. PHILBRICK, ALLEN K. *The Geography of Education in the Winnetka and Bridgeport Communities of Metropolitan Chicago*
*9. BRADLEY, VIRGINIA. *Functional Patterns in the Guadalupe Counties of the Edwards Plateau*
*10. HARRIS, CHAUNCY D., and FELLMANN, JEROME DONALD. *A Union List of Geographical Serials*
*11. DE MEIRLEIR, MARCEL J. *Manufactural Occupance in the West Central Area of Chicago*
*12. FELLMANN, JEROME DONALD. *Truck Transportation Patterns of Chicago*
*13. HOTCHKISS, WESLEY AKIN. *Areal Pattern of Religious Institutions in Cincinnati*
*14. HARPER, ROBERT ALEXANDER. *Recreational Occupance of the Moraine Lake Region of Northeastern Illinois and Southeastern Wisconsin*
*15. WHEELER, JESSE HARRISON, JR. *Land Use in Greenbrier County, West Virginia*
*16. MCGAUGH, MAURICE EDRON. *The Settlement of the Saginaw Basin*
*17. WATTERSON, ARTHUR WELDON. *Economy and Land Use Patterns of McLean County, Illinois*
*18. HORBALY, WILLIAM. *Agricultural Conditions in Czechoslovakia, 1950*
*19. GUEST, BUDDY ROSS. *Resource Use and Associated Problems in the Upper Cimarron Area*
*20. SORENSEN, CLARENCE WOODROW. *The Internal Structure of the Springfield, Illinois, Urbanized Area*
*21. MUNGER, EDWIN S. *Relational Patterns of Kampala, Uganda*
*22. KHALAF, JASSIM M. *The Water Resources of the Lower Colorado River Basin*
*23. GULICK, LUTHER H. *Rural Occupance in Utuado and Jayuya Municipios, Puerto Rico*
*24. TAAFFE, EDWARD JAMES. *The Air Passenger Hinterland of Chicago*
*25. KRAUSE, ANNEMARIE ELISABETH. *Mennonite Settlement in the Paraguayan Chaco*
*26. HAMMING, EDWARD. *The Port of Milwaukee*
*27. CRAMER, ROBERT ELI. *Manufacturing Structure of the Cicero District, Metropolitan Chicago*
*28. PIERSON, WILLIAM H. *The Geography of the Bellingham Lowland, Washington*
*29. WHITE, GILBERT F. *Human Adjustment to Floods: A Geographical Approach to the Flood Problem in the United States*
30. OSBORN, DAVID G. *Geographical Features of the Automation of Industry*   1953. 120 pp.
*31. THOMAN, RICHARD S. *The Changing Occupance Pattern of the Tri-State Area, Missouri, Kansas, and Oklahoma*
*32. ERICKSEN, SHELDON D. *Occupance in the Upper Deschutes Basin, Oregon*
*33. KENYON, JAMES B. *The Industrialization of the Skokie Area*
*34. PHILLIPS, PAUL GROUNDS. *The Hashemite Kingdom of Jordan: Prolegomena to a Technical Assistance Program*
*35. CARMIN, ROBERT LEIGHTON. *Anápolis, Brazil: Regional Capital of an Agricultural Frontier*
*36. GOLD, ROBERT N. *Manufacturing Structure and Pattern of the South Bend–Mishawaka Area*
*37. SISCO, PAUL HARDEMAN. *The Retail Function of Memphis*
*38. VAN DONGEN, IRENE S. *The British East African Transport Complex*
*39. FRIEDMANN, JOHN R. P. *The Spatial Structure of Economic Development in the Tennessee Valley*
*40. GROTEWOLD, ANDREAS. *Regional Changes in Corn Production in the United States from 1909 to 1949*
*41. BJORKLUND, E. M. *Focus on Adelaide—Functional Organization of the Adelaide Region, Australia*
*42. FORD, ROBERT N. *A Resource Use Analysis and Evaluation of the Everglades Agricultural Area*
*43. CHRISTENSEN, DAVID E. *Rural Occupance in Transition: Sumter and Lee Counties, Georgia*
*44. GUZMÁN, LOUIS E. *Farming and Farmlands in Panama*

    * Out of print.

*45. ZADROZNY, MITCHELL G. *Water Utilization in the Middle Mississippi Valley*
*46. AHMED, G. MUNIR. *Manufacturing Structure and Pattern of Waukegan–North Chicago*
*47. RANDALL, DARRELL. *Factors of Economic Development and the Okovango Delta*
48. BOXER, BARUCH. *Israeli Shipping and Foreign Trade*   1957. 176 pp.
*49. MAYER, HAROLD M. *The Port of Chicago and the St. Lawrence Seaway*
*50. PATTISON, WILLIAM D. *Beginnings of the American Rectangular Land Survey System, 1784–1800*
     1957. 2d printing 1963. 260 pp. Available from Ohio Historical Society.
*51. BROWN, ROBERT HAROLD. *Political Areal-Functional Organization: With Special Reference to St. Cloud, Minnesota.*
52. BEYER, JACQUELYN. *Integration of Grazing and Crop Agriculture: Resources Management Problems in the Uncompahgre Valley Irrigation Project.*
53. ACKERMAN, EDWARD A. *Geography as a Fundamental Research Discipline*   1958. 40 pp. $1.00
*54. AL-KHASHAB, WAFIQ HUSSAIN. *The Water Budget of the Tigris and Euphrates Basin*
55. LARIMORE, ANN EVANS. *The Alien Town: Patterns of Settlement in Busoga, Uganda*   1958. 210 pp.
56. MURPHY, FRANCIS C. *Regulating Flood-Plain Development*   1958. 216 pp.
*57. WHITE, GILBERT F., *et al. Changes in Urban Occupance of Flood Plains in the United States*
58. COLBY, MARY MC RAE. *The Geographic Structure of Southeastern North Carolina*
*59. MEGEE, MARY CATHERINE. *Monterrey, Mexico: Internal Patterns and External Relations*
60. WEBER, DICKINSON. *A Comparison of Two Oil City Business Centers (Odessa-Midland, Texas)*
     1958. 256 pp.
61. PLATT, ROBERT S. *Field Study in American Geography*   1959. 408 pp.
62. GINSBURG, NORTON, editor. *Essays on Geography and Economic Development*   1960. 196 pp.
63. HARRIS, CHAUNCY D., and FELLMANN, JEROME D. *International List of Geographical Serials*
     1960. 247 pp.
*64. TAAFFE, ROBERT N. *Rail Transportation and the Economic Development of Soviet Central Asia*
*65. SHEAFFER, JOHN R. *Flood Proofing: An Element in a Flood Damage Reduction Program*
*66. RODGERS, ALLAN L. *The Industrial Geography of the Port of Genova*
67. KENYON, JAMES B. *Industrial Localization and Metropolitan Growth: The Paterson-Passaic District.*   1960. 250 pp.
68. GINSBURG, NORTON. *An Atlas of Economic Development*
     1961. 119 pp. 14×8½″. Cloth. University of Chicago Press.
69. CHURCH, MARTHA. *Spatial Organization of Electric Power Territories in Massachusetts*
     1960. 200 pp.
70. WHITE, GILBERT F., *et al. Papers on Flood Problems*   1961. 234 pp.
71. GILBERT, E. W. *The University Town in England and West Germany*
     1961. 79 pp. 4 plates. 30 maps and diagrams.
72. BOXER, BARUCH. *Ocean Shipping in the Evolution of Hong Kong*   1961. 108 pp.
*73. ROBINSON, IRA M. *New Industrial Towns of Canada's Resource Frontier*
     (Research Paper No. 4, Program of Education and Research in Planning, The University of Chicago.)
74. TROTTER, JOHN E. *State Park System in Illinois*   1962. 152 pp.
75. BURTON, IAN. *Types of Agricultural Occupance of Flood Plains in the United States*
     1962. 167 pp.
*76. PRED, ALLAN. *The External Relations of Cities during 'Industrial Revolution'*
77. BARROWS, HARLAN H. *Lectures on the Historical Geography of the United States as Given in 1933*
     Edited by WILLIAM A. KOELSCH.   1962. 248 pp.
*78. KATES, ROBERT WILLIAM. *Hazard and Choice Perception in Flood Plain Management*
79. HUDSON, JAMES. *Irrigation Water Use in the Utah Valley, Utah*   1962. 249 pp.
*80. ZELINSKY, WILBUR. *A Bibliographic Guide to Population Geography*
*81. DRAINE, EDWIN H. *Import Traffic of Chicago and Its Hinterland*
*82. KOLARS, JOHN F. *Tradition, Season, and Change in a Turkish Village*
     NAS-NRC Foreign Field Research Program Report No. 15.
*83. WIKKRAMATILEKE, RUDOLPH. *Southeast Ceylon: Trends and Problems in Agricultural Settlement*
84. KANSKY, K. J. *Structure of Transportation Networks: Relationships between Network Geometry and Regional Characteristics*   1963. 155 pp.
*85. BERRY, BRIAN J. L. *Commercial Structure and Commercial Blight*
86. BERRY, BRIAN J. L., and TENNANT, ROBERT J. *Chicago Commercial Reference Handbook*
     1963. 278 pp.
*87. BERRY, BRIAN J. L., and HANKINS, THOMAS D. *A Bibliographic Guide to the Economic Regions of the United States*
*88. MARCUS, MELVIN G. *Climate-Glacier Studies in the Juneau Ice Field Region, Alaska*
89. SMOLE, WILLIAM J. *Owner-Cultivatorship in Middle Chile*   1964. 176 pp.
90. HELVIG, MAGNE. *Chicago's External Truck Movements: Spatial Interaction between the Chicago Area and Its Hinterland*

     * Out of print.

91. HILL, A. DAVID. *The Changing Landscape of a Mexican Municipio, Villa Las Rosas, Chiapas*
    NAS-NRC Foreign Field Research Program Report No. 26. 1964. 121 pp.
92. SIMMONS, JAMES W. *The Changing Pattern of Retail Location* 1964. 202 pp.
93. WHITE, GILBERT F. *Choice of Adjustment to Floods* 1964. 150 pp.
94. MCMANIS, DOUGLAS R. *The Initial Evaluation and Utilization of the Illinois Prairies, 1815–1840*
    1964. 109 pp.
95. PERLE, EUGENE D. *The Demand for Transportation: Regional and Commodity Studies in the
    United States* 1964. 130 pp.
*96. HARRIS, CHAUNCY D. *Annotated World List of Selected Current Geographical Serials in English*
97. BOWDEN, LEONARD W. *Diffusion of the Decision To Irrigate: Simulation of the Spread of a New
    Resource Management Practice in the Colorado Northern High Plains* 1965. 146 pp.
98. KATES, ROBERT W. *Industrial Flood Losses: Damage Estimation in the Lehigh Valley*
    1965. 76 pp.
99. RODER, WOLF. *The Sabi Valley Irrigation Projects* 1965. 213 pp.
100. SEWELL, W. R. DERRICK. *Water Management and Floods in the Fraser River Basin* 1965. 163 pp.
101. RAY, D. MICHAEL. *Market Potential and Economic Shadow: A Quantitative Analysis of Indus-
     trial Location in Southern Ontario* 1965. 164 pp.
102. AHMAD, QAZI. *Indian Cities: Characteristics and Correlates* 1965. 184 pp.
103. BARNUM, H. GARDINER. *Market Centers and Hinterlands in Baden-Württemberg* 1966. 172 pp.
104. SIMMONS, JAMES W. *Toronto's Changing Retail Complex* 1966. 126 pp.
105. SEWELL, W. R. DERRICK, et al. *Human Dimensions of Weather Modification* 1966. 423 pp.
106. SAARINEN, THOMAS FREDERICK. *Perception of the Drought Hazard on the Great Plains* 1966 .183 pp.
107. SOLZMAN, DAVID M. *Waterway Industrial Sites: A Chicago Case Study* 1967. 138 pp.
108. KASPERSON, ROGER E. *The Dodecanese: Diversity and Unity in Island Politics* 1967. 184 pp.
109. LOWENTHAL, DAVID, editor. *Environmental Perception and Behavior* 1967. 88 pp.
110. REED, WALLACE E. *Areal Interaction in India: Commodity Flows of the Bengal-Bihar Industrial
     Area* 1967. 210 pp.
*111. BERRY, BRIAN J. L. *Essays on Commodity Flows and the Spatial Structure of the Indian Economy*
112. BOURNE, LARRY S. *Private Redevelopment of the Central City, Spatial Processes of Structural
     Change in the City of Toronto* 1967. 199 pp.
113. BRUSH, JOHN E., and GAUTHIER, HOWARD L., JR. *Service Centers and Consumer Trips: Studies
     on the Philadelphia Metropolitan Fringe* 1968. 182 pp.
114. CLARKSON, JAMES D. *The Cultural Ecology of a Chinese Village, Cameron Highlands, Malaysia*
     1968. 174 pp.
115. BURTON, IAN, KATES, ROBERT W., and SNEAD, RODMAN E. *The Human Ecology of Coastal Flood
     Hazard in Megalopolis* 1968. 196 pp.
*116. MURDIE, ROBERT, *Factorial Ecology of Metropolitan Toronto, 1951–1961*
117. WONG, SHUE TUCK, *Perception of Choice and Factors Affecting Industrial Water Supply Deci-
     sions in Northeastern Illinois* 1968. 96 pp.
118. JOHNSON, DOUGLAS. *The Nature of Nomadism: A Comparative Study of Pastoral Migrations
     in Southwestern Asia and Northern Africa* 1969. 200 pp.
119. DIENES, LESLIE. *Locational Factors and Locational Developments in the Soviet Chemical Industry*
     1969. 285 pp.
120. MIHELIC, DUSAN. *The Political Element in the Port Geography of Trieste* 1969. 104 pp.
121. BAUMANN, DUANE. *The Recreational Use of Domestic Water Supply Reservoir: Perception and
     Choice* 1969. 125 pp.
122. LIND, AULIS O. *Coastal Landforms of Cat Island, Bahamas: A Study of Holocene Accretionary
     Topography and Sea-Level Change* 1969. 156 pp.
123. WHITNEY, JOSEPH. *China: Area, Administration and Nation Building* 1970. 198 pp.
124. EARICKSON, ROBERT. *The Spatial Behavior of Hospital Patients: A Behavioral Approach to Spatial
     Interaction in Metropolitan Chicago.* 1970. 198 pp.
125. DAY, JOHN CHADWICK. *Managing the Lower Rio Grande: An Experience in International River
     Development.* 1970. 277 pp.
126. MACIVER, IAN. *Urban Water Supply Alternatives: Perception and Choice in the Grand Basin On-
     tario.* 1970. 178 pp.
127. GOHEEN, PETER G., *Victorian Toronto, 1850 to 1900: Pattern and Process of Growth* 1970. 278 pp.
128. GOOD, CHARLES M. *Rural Markets and Trade in East Africa* 1970. 252 pp.
129. MEYER, DAVID R. *Spatial Variation of Black Urban Households* 1970. 127 pp.
130. GLADFELTER, BRUCE, *Meseta and Campiña Landforms in Central Spain* 1971. 204 pp.
131. MOLINE, NORMAN, *Mobility and the Small Town, 1900–1930* 1971
132. NEILS, ELAINE, *Reservation to City: Indian Urbanization and Federal Relocation* 1971
133. SCHWIND, PAUL, J. *Migration and Regional Development in the United States* 1971. 170 pp.